Jump Start
Your Career

Finding the Right Job after Military Service

♦ Colonel Jerry Crews ♦
U.S. Army Retired
Military Officers Association of America

THOMSON

™

CUSTOM PUBLISHING

Editor: Tim Spurlock
Director of Product Creation: Becky Schwartz
Manufacturing Supervisor: Donna M. Brown
Pre-Media Services Supervisor: Christina Smith
Graphic Designer: Krista Pierson
Rights and Permissions Specialist: Kalina Ingham Hintz
Project Coordinator: Jennifer Flinchpaugh
Marketing Manager: Sara L. Hinckley

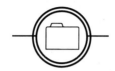

Contents

INTRODUCTION

My Background for Writing this Book

robably the first question you might have about this book is, "what makes this guy so smart that he thinks his book is worth reading?" Well, honestly, as you will quickly find out, I am not that smart but I have been around the world a few times, went to a rodeo in Texas, hiked most of Mt Rainier in Washington state and was shot down in Laos many years ago. So, how does any of that qualify me to write about the career transition business? Well it probably doesn't. But with over 29 years of military service, nine years at the Military Officers Association of America –lecturing, writing and counseling others about their career transitions– and a few more years of worldly experience, I've learned a lot about life, reinforced many other things I already knew, and now have some of my own philosophy worth sharing. In fact, I now know there are a number of roads I went down simply to find that I didn't need to go down the road in the first place. Also, there are many, many books that I don't need to read because I already know the ending—this is what you get out of life at my age. Further, I've found there are very few original thoughts left in life and, of course, you will not find many in this book. More importantly, most of what you read in this book will be what others have shared with me based on their trials and tribulations of life.

As for me, I grew up poor—probably will die poor—worked on a tobacco plantation in the summers, gas stations in the winter, delivered newspapers, and acquired a host of other street savvy "skills" that allowed me to survive in Norfolk, Virginia with a little spending money. Upon graduating from high school, with no particular knowledge or direction in life, I took six years to get a degree in Sociology from Old Dominion College (now University), and then joined the Army based on guidance I received from a Judge in Virginia Beach, Virginia. All in all not a spectacular start in life, and like many of you reading this book, I just stumbled along and one day life found me in the Army as a private. For some very odd reasons, the Army –after basic training– began to appeal to me. As I looked around and saw the opportunities to travel, jump out of airplanes, do things I had never even thought of doing, my interest in life began to change; plus, I really didn't have any other callings. The Army offered me the opportunities to become a paratrooper, Ranger, Pathfinder, Helicopter Pilot, I Served in Special Forces, Armored and Infantry Divisions, invested five years of my life in the Pentagon, got a Masters Degree and attended the National War College. A "hoorah" soldier was I. I received a number of awards for two tours of duty in the Republic of Vietnam, all of which has very little to do with my career transition business but a lot about forming some of the "philosophy" I will weave throughout this book.

I've always been lucky and had an uncanny ability to "fall into" great opportunities. A retired Army warrant officer, Don Hess, told me about an opportunity at The Retired Officers Association; now the Military Officers Association of America. The opportunity allowed me to continue my association with, and service to, the entire military community. The position was as a deputy director in The Retired Officers Association (TROA), Officer Placement Service (TOPS) Directorate, writing, counseling, and lecturing to military personnel transitioning from the military to the civilian world. A retired Navy admiral didn't initially appreciate what a great American soldier I was on my first interview but later came to his senses based on the wise counsel of another retired Army warrant officer, Bob Johnston, who worked for him. For the last nine years, I have traveled all over the world giving lectures, writing monthly career transition articles, and counseling officers about the "challenges" they will face during their career transition to the civilian world. In this book, I will share many of those "lessons of life" and personal philosophy of life. I will not sugar coat anything. This book is titled "Jump Start" for a reason. It is intended to get you off the starting blocks quickly. It is not a narrative about life. You can read about that some other time. Right now you need a job or you wouldn't be reading this book. So, start reading!

What You Can Expect to Get from this Book

On reflecting upon my many years on this planet, it dawned on me that many of the lessons I've learned in life, both mine as well as others, came about through the "school of hard knocks." Many of the books I've enjoyed were short, to the point, filled with humor and relevant "war stories." I've been around the world a few times and in the career transition business for over nine years. I have come to the conclusion that the basic principles of finding a job, career transitions, and related challenges associated with today's working world are all the same. This book is generally written for the military person who has served a tour of active duty, or more likely is retiring after a twenty or thirty year career of military service. The key principles of job-hunting, during career changes, are all the same regardless of whether you are a civilian or ex military person. Linemen on a football team all have the same basic stance and strive to keep the opposing linemen from getting through their line. The difference on each team is that some are faster, some are bigger, and some have quicker instincts about the game but the basics of their game are all the same.

Thousand of books can be found in any book store or library on how to "find yourself," write a resumé, network, chase the money, don't chase the money but chase life, dress, get your foot in the door, interview, and on and on. You name it and someone has written a book about it—and written it with far more eloquence than I will. My intent in this short book is to write something that will get you immediately launched into the job-hunting process and you can then add or change courses as you go. Many people, both in the military as well as the civilian–world, want to have everything in place– all the "T's" crossed and the "I's" dotted before they move out to find a job. That is time wasted. The intent of this book is to address the everyday person, like me, who came on this planet poor, joined the military because they were poor and will most likely be placed in their graves poor. This book is not about wealthy people who may have a covey of lawyers helping them negotiate stock options, buy outs, severance packages and a host of executive privileges that we will never know of in our working lives. This book is about your everyday, hard working middle- class, blue collar person, who has those core values of honesty, hard work for a days pay, self discipline of never missing a day's work, task orientation not time orientation, and loyalty to their boss as well as the organization.

While I would suggest reading a few transition books (maybe two), you will find that– with few exceptions– the general approach for all is the same, i.e., decide what you want or can do, write a resumé, find a job, apply for a it, get interviewed, get hired, and go to work. The intent of my book is, yes, to cover many of these "basics" but to also write it in everyday language, with some philosophy woven throughout, that you can relate to as you put your individual plan together. I would hope that you would immediately decide that chasing money and material things is not the answer to changing careers. I will also note that if you are retiring from the military you have probably had your one and only career. The rest of your life you are simply going to get a job and work, work, work; then, one day you are going to die. Oh, by the way, some of the words you will find in the definition of work are, toil, labor, drudgery—not travel, fun, education, camaraderie.

After reading this book, hopefully, you will come up with a plan for your life that includes a balanced approach to working in a field that not only challenges you but also offers the opportunity to pursue a quality of life that enriches you and your family. That old joke "life is short, eat dessert first" has more meaning every day you wake up. Your career plan should include enough money to live a fulfilling and challenging life, which includes being healthy and happy, until the day you are buried with the possibility that your check for the casket may bounce. If your plan is executed to perfection, your children will not have to borrow money to bury you–but if they do–so be it, you borrowed money to raise them, send them through college, and get them married. They do owe you a decent burial.

The Decision to Change Careers

The decision to leave the military comes about in one or two ways. You make the decision or it's made for you by the military. Generally, if you make the decision, you feel much better about the entire separation than if it's made for you when you have not been selected for promotion to the next higher grade. Regardless, of how or why you leave the military, you must do so with a good attitude. Negative attitudes only prolong the separation anxiety and turn others against wanting to help you. Life is not fair. Some people get promoted deservingly, others just get promoted, and a few others probably should never have been promoted. Again, life is not fair. You can't fix it, so don't dwell on it. You'll have enough challenges that you can fix. Get on with your life.

One of the first dilemmas you must cope with if you want to transition successfully is loyalty. There has been no question about your loyalty; it has been instilled in you from the first day you put on your uniform. You have been loyal to your command, your commander, and those who work for you. You are so loyal you'll work 18 hours a day until your last day of service, when you're presented your final award and hang up your uniform for the last time. But in today's highly competitive job market you cannot remain so loyal that you wait until the last day of service to begin your job hunt. You must start preparing for separation, or retirement, at least a year out and sooner if possible.

Many of you believe your job is so important that only you can do it, but don't use this excuse to avoid dealing with the civilian world—everyone can be replaced. And, if you think you are irreplaceable, you may have a larger problem that will effect your career change—like an ego that others can't accept. You owe loyalty not only to your bosses but also to your family. Show them that by devoting time, resources, and attention to having a job after you leave the military. After all, living indoors and eating regularly are habits we all have come to enjoy.

Five "Keys" to Happiness

As you think through a plan for you and your family about what you want out of life after military service, I want to share with you a "filter" you should use as your guide for a more rewarding life. The five keys to happiness are not necessarily discussed in priority because they all are interdependent and need to be balanced if you are to enjoy the richness and fullness of a balanced life. Most importantly, short of the good Lord becoming involved, you have almost absolute control over these.

1. Health: Probably the most important thing in your life is good health. The three aspects of good health that you can control are smoking, weight, and drinking. If you smoke, it's the dumbest thing you do and you know it. You can quit smoking! You will quit smoking one day; unfortunately, you will not do it pleasantly. No, you will have lung cancer, throat cancer, emphysema, and a host of other smoke-related ills that will cause you to drag your family and friends through a prolonged series of medical crises and financial hardships before you finally die. And, when you finally die, you will be glad the suffering is over. If you don't want to quit for health and family reasons, then you might want to consider this fact of life. Given a choice, people who don't smoke will not hire those who do because they have found that smokers are sick more often than other employees, smell like smoked salmon, and want a 15 minute smoke break every 10 minutes.

The next factor of health you control is weight. There are only two reasons most people are over weigh–they eat too much and don't exercise enough. Medically, many argue that being overweight is more harmful than smoking. You don't need fancy diets to keep your weight under control. The average male requires about 1800 calories a day and the average female about 1400 calories. How you get them is not as important as staying within the limits. Throw in a daily walk or other relevant exercise and you will maintain your weight.

Finally, it is easy to drift from a few drinks a day to a few bottles a day. You all know the hazard of excessive alcohol abuse; I will not take your time to beat this one to death. I will say that any of the three problems I've discussed fall under one word—discipline.

2. Love: It took me a divorce, remarriage, and many years to discover how important love is. No, naval aviators, not love of yourself, but love of someone else. If you give love to others, it's like karma and will come back to you in ways you can never imagine. If you stay angry and carry negative feelings, you will eventually be the only one effected because others will simply ignore you and go their own ways. Learn to give as well as respond to love and remember that sometimes you really have to work at it to be successful. If you are willing to give with all your heart you will be rewarded with happiness that I'm unable to express in this book.

3. Financial Security: Many of you reading this book are deep in debt because you are trying to keep up with the Jones' and are buying things you have no need for and, in many cases, don't even want. Some of you have china, crystal and silver sets that have never had a human lip touch them, five bedroom, four and half bathroom homes that only one guest has ever slept in—Why? Do you need a credit card for every specialty store? If you have more than one credit card and can't pay it off at the end of the month, you have problems. With large debts you trap yourself into seeking jobs for the money and not the quality of life. Get your spending under control. Do not give your children money. Let them earn it the same way you did. My long-range plan is to spend every penny I have and my children will have to borrow money to bury me.

4. Spirituality: While I'm not a regular church member or demonstrate great religious convictions, I do believe there is a larger "being" that has impact on our destiny. You cannot sit on your porch and watch birds build a nest, watch a nursing animal, or have a small child hold your hand and not know that there must be something larger than you that makes this occur. If you lack this sense of spiritual understanding and well being, then you are destined for loneliness when you are in a time of need.

5. Second Career: Your work generally consumes more of your life than any other endeavor. It defines who you are–in many cases your values–and of course, provides the financial platform for your life. You must be the one who defines what a successful job or career is not your neighbor, friends, or the "Joneses" of life. If you want to drive an 18-wheeler logging truck, and listen to Merle Haggard and Johnny Cash's greatest hits and it makes you happy, then go for it. Unless you win a lottery, you are most likely going to die poor so you may as well be happy doing what you do in your work.

Now, that I've covered the five most important things in life, I want to share some deeper and more philosophical "Pearls of Wisdom" that you can use as a filter when you and the family are focusing on your, and their, career changes.

What's Really Important in Your Second Career?

I'm not sure at what point in our lives we become worth listening to, but I've been on this planet for more than 50 years and want to share some thoughts on second careers. There are very few original thoughts left about life, liberty, and the pursuit of happiness, so much of what I have to say will not be original but I hope it will prompt you to think hard about what's important in your life as you seek out fame, fortune, or serenity in your new job track.

When I'm not on the road traveling and lecturing, I'm in my office counseling officers of all the uniformed services on how to be "successful" in their second careers. I admit that when I first started counseling officers on which road to take after taking off their uniform for the last time, I usually identified money as the deciding factor. The longer I am in this business, the less importance I attach to money and material things and the more importance I attach to lifestyles and happiness.

I certainly would not suggest we all surrender our worldly possessions and join a monastery. But as we seek to maintain a comfortable lifestyle, we also should keep some perspective on what's truly important in our lives. The tales of woe I hear as a counselor are not concerned solely with money but with the lack of personal and family time, and with other aspects of the work environment. In many cases retired officers are frustrated by their colleagues, office conditions, a perceived loss of status, a feeling that they lack control of their destiny, and other small but important aspects of their lives. Most of the officers I speak with have difficulty answering two questions: "What do I want to do?" and "How much does it pay?" Without a doubt, the first question is more important. The answer will eventually provide you happiness—but not necessarily money!

When officers tell me they have to have a certain amount of money, I ask them why they need so much and what they are going to do with it if they get it. More often than not, they respond with a stare that suggests I must be crazy. Then I ask them, "How much food can you eat? How many cars do you need?" I want them to consider what they are going to do with money if they have no time to enjoy it with family and friends. If you are around 40 years old and haven't owned a Mercedes yet, do you really need one to make you happy? Secretly, I would like to have a 320-E, but my life will not change without it. If I owned a Mercedes I would have to eat less to make the payments and worry constantly about it being stolen, adding more stress to my life.

Most officers who have reached lieutenant colonel or commander and above will never make as much money after military service as they made on active duty, no matter how hard they work at it. We've all heard stories about a retired officer making six figures, but that is extremely rare, and in most cases those people are working longer hours under more stressful conditions than they did on their worst days at the Pentagon.

So, let's go back to the basic questions and how to answer them. Most of you have high mortgages, children in school, car payments, and a host of smaller bills. The salary required to maintain the standard of living you enjoyed on active duty will vary depending on where you live. Life is a matter of trade-offs: Do you trade money for material things? Do you trade money for time with family and friends? Do you trade money for less stress? Consider that you've already climbed the military ladder. Do you really need to climb the corporate ladder in a second career?

This is why I now counsel officers to think of more than money when they plan their lives after the military. What are your personal goals, your family goals, your long term financial goals? What about the "fourth quarter of life" (i.e., life after the second career)? Sit down with your family and discuss what's important for your happiness and for theirs. Plan a strategy to get there long before you take your uniform off for the last time. There is far more to changing careers than money. You need to give those other considerations a lot of thought.

Keeping Up with the Jones

If you are trying to keep up with the Jones of life you are sitting at the kitchen table trying to figure out not only why you bought so much stuff you didn't need, but also–and more importantly–how you're going to pay for it. If you are trying to keep up with the neighbors down the street you may have "maxed out" your credit cards and will be doing well to have everything paid off by late summer— just in time to pay for your summer vacation, buy some books and clothes for children heading off to school, and begin some early Christmas shopping. Then the cycle will start all over again.

How many of you reading this article have "been there, done that" over the last few years? Does there seem to be no end to this continual bankruptcy of your life? There can be if you decide to take control of your and your family's future.

In my "Marketing Yourself for a Second Career" lectures, I often advise that if you have more than one credit card and more than $1,000 in credit card debt, you are headed for personal financial trouble. In fact, the average person has three credit cards and roughly $5,000 to $8,000 of credit card debt. (This debt is over and above the average family's home mortgage, car payment, grocery tab, and health care bill.)

If this sounds like you, you probably acquired and kept this debt by continually buying things that you didn't need in order to keep up with the Jones. Depending on when you charge an item to your credit card, how much unpaid balance you carry from month to month, and a few other hidden credit card factors, you easily could be paying more than $300 a month in interest on a $5,000 credit card bill.

I share these numbers with you not only as a wake-up call about your finances but also as a reminder to think about your priorities when you start planning the transition from your life in uniform to your start in the civilian world.

Large debts or an ambitious standard of living can make salary a disproportionately decisive factor in your career search, which can affect your ability to freely negotiate with a potential employer or narrow your career choices.

When I lecture, I frequently pose these questions, "How much can you eat?" "How many Hummel or Lladro figurines can you store in a cabinet?" "How many Waterford glasses do you need to entertain your good friends?" "Do you really need a five bedroom home for you and your spouse?" Finally, "If the Joneses are not paying your bills, why are you trying to keep up with them?" Behind these rhetorical questions, there are real issues to consider, which affect not only your long-term plans (can you afford to pursue the career you really want?) but also your short-term ones (will you make that trip to the mall this weekend?). If you have your health, do you really need a lot of money and material possessions to make you happy?

Once you've made the decision to retire from the service, and you're looking for a second career. Your options are already limited enough, so be sure that your hands aren't tied by your family's overspending now. Although you probably won't be able to match your current salary immediately when you retire; you should not have to lower your standard of living just because you are trading your uniform for a civilian suit. You also probably do not want your decision to accept or decline an offer to be based solely on the size of the salary, although this is clearly an important factor in whether or not you would accept a job.

If during an initial interview, an employer offers you a low-paying, mid-level entry position that has great growth potential, what would be your answer? Assuming the position is right for you, your answer should be, "I'll take it." However, if you have been out pushing the shopping cart around, buying things you don't need and acquiring high-interest debt, you may be forced to value money over other negotiable benefits simply to keep paying the bills. You don't want to have the burden of debt determine whether or not you take a job.

To avoid just such a situation, sit down with your entire family 12 to 18 months before you depart the service, and discuss your financial situation as it relates to the transition process. How much is your retired take-home pay and your spouse's salary combined? How much are your monthly bills? Are college expenses about to take a big bite out of your money? How much do you want to save or invest for the future, and where do you, as a family, want to be financially in the next five to 10 years? If, by the time you read this article, you have paid off all of your bills from last Christmas, you probably don't need to read further. If not, keep reading!

Next time you are in a military exchange, or any department store with shopping carts, look around, and watch what is happening. In many cases, there is a frenzy to fill the shopping cart. In some cases, entire families are throwing items of clothing, kitchen utensils, food, and assorted electronic devices in their carts. Some don't have the slightest idea what these items will be used for when they get home; need is not always the reason for throwing an item in the shopping cart. Sometimes their rationale will be, "I'll try it on at home to see if it fits; I can always bring it back." Or, "It has additional features that the last two we bought don't have." Or finally, "It's on sale and a great buy," even though you already have the exact same thing at home. If any of this sounds familiar, then you are adversely affecting your flexibility during your job search.

A recent article in *The Washington Post* stated that between one-half and two-thirds of all households carry an average $7,000 to $8,000 balance on their credit cards. That equals about $1,000 a year in interest payments, or almost $100 a month given to a lender at an average rate of 18.9 percent. The only difference between this and highway robbery is that credit cards are legal.

Many of those in the lending industry say that 96 percent of accounts are paid on time; I read this to mean that the cardholder is paying the interest on time, not necessarily the principal—which helps everyone in the financial community except the borrower. Unfortunately, few realize the adverse effect this "plastic disease" can have on your career options during your transition to the civilian world.

The bottom line is, if you lack financial discipline, you need to push the shopping cart over to the corner of the store and walk slowly—if necessary, run—away from the cart as though the building were on fire. Use a hand basket when you shop, and buy only what will fit into it.

Most of you have successfully completed one career and worked hard for what you have. Keep your bills paid, and save a little. Once you have begun to free yourself from debt, you can truly say to a potential employer that you want to study his offer, along with any other opportunities you may have. Your career decision will not be driven by what you have been throwing into the shopping cart. Let your next career choice be driven by more than money, and give yourself the luxury of taking any job you want.

I have been in this transition business, both in the military and as a civilian, for many years, and more than once I have heard a male officer say, "My wife just does not understand the problems I'm having finding a job, and it's affecting our family and marriage." I have never heard this complaint from a female officer. In one instance, I heard these words from a male officer who was a service academy graduate, held a master's degree from a prestigious university, had earned many military awards, and enjoyed some of the most career enhancing assignments during his service. So what was his problem? The biggest hurdle families face during this time is a lack of sensitivity and communication among themselves. Wives don't understand how their husbands can be so successful in the service and yet be nervous about their abilities as civilians. Husbands typically fall into one of two categories: those who have planned and those who have not planned their post-retirement years. Regardless of the category in which you fall, share your feelings and plans with your family.

Most officers do not know the details of their retirement because transitioning is territory they have not previously traveled. Nevertheless, even the not knowing should be talked about, and who better to discuss it with? Your family is the most affected by this change in your life. You may be surprised to find that your wife and young teenagers have feelings, concerns, and interests in what is happening to them. As a service member, you wouldn't think about going out on a tactical mission without briefing the team on every detail of the operation. Show your family the same respect.

Maybe it's the nature of military service, but male officers are often reluctant to show emotion and tend to keep personal feelings inside. During their transition, some men experience mood swings, outbursts of temper, or even psychosomatic physical illness and never once share it with their wives. Some think it is a sign of weakness. They are wrong! There is nothing inappropriate about showing emotion. If you are frustrated, share it with your family, and don't be surprised if they are warm and understanding. Whether you are bothered by career changes, lack of promotion, failure to find the "right" job, or differences in the civilian world, discuss it with your family. Your children will learn that sometimes life gets tough, even for old Dad, who has done so well in uniform. But don't dwell on the negatives. Keep a positive attitude and be upbeat, even when things are not going your way. This sets a great example.

If your wife is one of those unsung heroes who raised the family, was involved with everything on the base, and socially supported you as you moved through your military career, but has not had a civilian job in many years, encourage her to attend transition programs with you. She needs to understand the challenges you will soon face; she can't understand if she simply depends on you to tell her what's happening. The longer your wife has not worked as a civilian, the more important it is that she attend transition programs, even if you have to coax her. Attending these together also will help the flow of communication and pay great dividends during your transition.

Finally, many officers say, "If my wife had a job, it sure would take a lot off pressure off me to get a high paying job, and I could really look at what I want to do." This is a decision only you and your wife can make. I will suggest that many wives who have been out of the workforce for a long time hesitate to go back for the same reasons you may be having problems during your transition—they don't understand the civilian workforce and lack confidence in their abilities to survive there. Encouraging them to attend your transition programs will help allay their fears, as will sharing some of your challenges with them. If the kids are off to college, and the house is empty, you might be better off emotionally and financially if you both work; remember your grandmother saying "Idle minds are hands of the devil." If nothing else, it will give you something to talk and laugh about at dinner—that military life has prepared you to compete, and compete well, with civilians. (I don't want to get into how much better military folks compete; that's another article.)

Your Initial Transition Plan

You've never done it before, and it might be the most difficult project you ever undertake; But, your transition plan is also the most important document you and your family will ever put together.

Many times in your career you have received permanent change of station (PCS) orders and had to rough out a schedule of events for getting you and your family to your next duty station on time. The planning process for a change of career is similar but much, much harder.

Unlike your PCS plan, your career transition plan is your sole responsibility. You'll have no senior boss looking over your shoulder to see if you are doing it right and no support staff to do your typing and make your calls. You'll have no one telling you when, where, and how to get to your next assignment or what you'll be doing once you get there. Yet you must approach your transition to a civilian career with the same dedication and attention to detail that you would if a senior flag officer had given you this mission. To succeed, you must accomplish the following:

Identify available resources; Formulate a flexible plan, Execute your plan.

Preparing early is critical. Before you sit down to create a transition plan, read about career transitioning in the Military Officers Association of America (MOAA) booklet, "Marketing Yourself for a Second Career," available free to MOAA members. Many other career resources are available in local libraries and bookstores as well as in MOAA Officer Placement Service's library in Alexandria, Va. You need read only one or two of these books to get a basic sense of what you are about to do and why it is important.

The basics of career transition have rarely changed over the years; understanding how to write a resumé and cover letter, researching, networking, dressing, interviewing, and negotiating will be the key elements of the process. You will quickly find that there is no perfect plan. Do your reading at least one to five years before your proposed retirement date, as things like saving money, obtaining more and current technical education, and effectively developing a networking system require long lead times.

After getting a feel for the journey you are about to take, focus on what you want to do and where you want to do it. Until you know this, all other career decisions will be in limbo. You cannot write a focused resumé, get the right education, or target organizations and networking contacts without knowing what you want to do. Do not get behind the power curve, always playing catch-up, and waiting until the last minute to start preparing for your next career.

Many people who wait until the last minute to prepare for their transition end up taking any job that comes along and then move from job to job trying to find the right one. Attempt to know what you want to pursue before you start the job search! And by all means involve your family, for your transition effects them profoundly.

After you decide what you want to do, you can develop a networking system and lay out a campaign plan. This plan, in military operational terms, should be developed using the "reverse planning sequence." Figuratively speaking, you "put a line in the sand" and work backward from that date to your current date. The closer you are to your "E" (exit or employment) date, the more detailed the plan should be. It could be your last day of active duty or the first day an employer can expect you to come to work. Without knowing the exact date of availability for a job, you cannot develop a detailed plan of execution. A year is about the minimum time required to execute a detailed career transition, assuming the other elements of your campaign—education, finances, and family considerations—come together.

The plan must have specific milestones expressed in terms of weeks, or other achievable goals, such as completing an advanced degree, children graduating from high school, or having a certain amount of money saved. Without realistic, achievable goals, you will simply become derailed at key times in the plan. Also, your plan should be flexible, for you will undoubtedly have to modify parts of it as you go along.

Write your plan as you would write a plan for any other important program, "staff" it with the entire family, and periodically review your program to make sure you stay on course. Finally, share this plan with your bosses and ask for their support early on so they will understand why you need time off for your service transition program, visits to job fairs, and many other things that will require time away from work. Do not wait until your last week of active duty to start planning your next career, or you may be unemployed for a long time.

The Basics

Most books on career transitioning claim to have some new approach on how to be successful in your job search. For $25 to $40 you can buy a book that will explain, step by step, how to successfully find a job. There is nothing wrong with most of these books, but why pay for information available free at the library, from your base transition office, or on the Internet? Read one or two books, then get on with some other phase of your career transition process e.g., networking or informational interviewing. A retired Colonel once wrote me a note saying "There are no secrets to finding a job beyond having a focused, organized, and disciplined approach that will make you successful. I set up a three-ring binder, wrote an 'operations order,' made contacts with people in the career field I wanted to pursue, and joined a professional organization. This kept me on track and organized."

He stuck to the basics, and in three months, he was working in his chosen career field. His job search included a focused cover letter accompanying each tailored resumé he sent to prospective employers. From the 70 targeted letters he sent out, he received six interviews (about par for that number of letters) leading to five job offers.

You do not need a one-of-a-kind resumé to successfully find a job. Find the resumé of someone in your field of interest who has similar experience and education, change the key pieces of information that require it, and you have a resumé. Some people spend a lot of time reading books, writing their resumés, and rewriting them, and eventually develop a severe case of "analysis paralysis." Instead, they should be out shaking hands and getting involved in social and professional activities. Successful job seekers say they were focused, organized, worked hard, et-cetera, but finally got a job "out of the blue" through a networking contact.

Let me highlight some of the "basics" to a successful career transition. These are starting points to put you well on your way to finding a satisfying second career.

- *Know what you want to do, when you want to do it, and where you want to do it.* You must have this information before you can begin to focus your time and energies in the right direction.
- *Networking is possibly the most important element of a successful career transition.* Once you have decided what, when, and where you are going to pursue a second career, the key to opening doors in your field will be someone already in that field.
- *Understand how your skills, education, experience, and personal goals fit today's job market.* Being a rifle platoon leader or Army Ranger is not as relevant as knowing about information management, logistics, health care, intelligence, security or contracting.
- *Write your "operations order," and share it with your entire family* or better yet, have your family participate in planning the future and then share it with your boss.
- *Write a simple, focused resumé* on one or two pages of plain bond paper and write in words that potential employers will understand. Your resumé should whet their appetite to call you for an interview. Do not write a lengthy "obituary."

Commitment to your new career is as important as accomplishing a mission for your military boss and deserves the same effort, enthusiasm, and dedication. Remember, if you fail a military mission, the people in your command suffer; if you are not successful in your second career, your family suffers. In either case, "failure to plan is planning to fail."

1

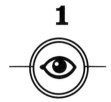

Perceptions of the Military

"Military people, by the very nature of what they do, have never had to look for work, nor do they really know how to work." Think about this statement and you will see why military members need help when transitioning to the civilian world after service in uniform. Some civilian employers think the entire military establishment does one of two things: They wreak havoc on other warriors (or in more recent years, keep warriors from attacking themselves or their fellow warriors under the guise of "peacekeeping" missions), or they train to wreak havoc (keep the peace). Regardless of which role military people pursue, it is a full time endeavor, and they never had time to pursue a real job. Combating misperceptions is a real problem for military people who step out into the world of civilian work after three to thirty years of service. The people who have been served by the military community often know very little about the skills military people can bring to an organization. Let's put it another way; many civilians don't understand what it is we weirdos do for a living!

Some civilians, for example, think that military people are not very flexible. Yet most have lived in two or three foreign countries, traveled or had temporary duty in five or six others, and lived in ten or fifteen states. Could any civilian on the street sell or rent their home as easily, get their children out of one school and enrolled in another, convince a spouse to quit another job, and make arrangements to ship their family, car, and household goods to a non-English speaking country with 30 days notice? Military people often arrive at their next assignment in just enough time to take off on another warrior or training mission, leaving the rest of the family to establish yet another home by themselves. Inflexible? Hardly.

Civilian employers can also think that former military members have weird work habits. Military work ethics are not always understood or found in the civilian community. Every day of their military working lives, service members come to work on time, in the right uniform, with a clear head, physically fit, and full of enthusiasm. They are loyal and will work 18 hours a day, seven days a week, to complete a project. And if the forecast is for bad weather, they will get up earlier just to be there on time. Perhaps anyone who would do that is truly weird and civilian employers are worried that former military members just won't fit in. Right?

Guess what else civilian employers don't often realize? Every officer and many non-commissioned officers have at least a bachelor's degree, most mid-level officers have master's degrees, and many hold doctorates in every discipline from music to nuclear physics, from bush league to Ivy League schools. Why would someone with that kind of education want to stay in the military? It's no wonder military people need help, for they must not only learn about the civilian work community but also learn to "market" themselves in words and actions that many civilians don't understand, and in some cases, don't appreciate. Military folks might need to get out and tell civilian employers how our weirdness can benefit their organizations.

We need to talk to them in terms they understand, dress like they do, even show them we can accomplish projects on time, ahead of schedule, and improve their organizations. We need to help ourselves by doing research, reading business journals and newspapers, and getting involved in civilian clubs and organizations. Next thing you know, we won't seem so weird, we will have become civilians.

2

The Plan: Preparing Yourself and Your Family

Anxious about Leaving the Military?

Unless you're blessed with the serenity of Buddha, the prospect of hanging up your uniform for good could be one of the most traumatic events of your life.

Anxieties are a natural part of the career transition process, whether you make the decision to leave the uniformed services or the decision is made for you by a Selective Early Retirement Board (SERB) or some other personnel action. Fortunately, planning ahead can diminish those anxieties whenever and however your retirement occurs.

Involuntary separation by the way, should not affect your successful transition to a civilian career. Officers involuntarily leaving today's uniformed services have no reason to feel embarrassed or feel less-than-worthy. Their situation is not unlike an Olympic skier competing for a gold medal. Every skier expects a chance for the gold medal, but the difference between gold and tears is often one one-hundredth of a second on the downhill slalom run. That's how close the call can be on selection boards that decide which officers stay and which ones must leave involuntarily. Moreover, if a potential employer is more interested in whether you were involuntarily released from the service or how much retired pay you get than in the skills you can bring to a job, then you don't want to work for him or her anyway.

Anxiety is highest when a difficult decision has to be made, and the longer you put off the decision, the greater the anxiety. The way to address anxiety is to gather as much information as possible, develop a plan, make a decision, and then get on with executing the decision. This strategy appears straightforward but I interview officers everyday who know they have a mandatory retirement date or will be changing careers and who have yet to come to grips with what they will do, where they want to do it, and how to go about planning for the transition.

Everyone turns in their uniform one day, so you can reduce the anxiety associated with career changes by long-term retirement planning. This planning includes investments, continuing-education, keeping abreast of political and economic changes, becoming and staying involved with people and organizations in work and hobby fields, and generally being aware of the ever-changing world scene. The downsizing of the defense community and increase in defense associated industries should not be news to anyone. But many officers simply are not prepared mentally, financially, or educationally for the civilian corporate community they must compete in for a successful second career.

The anxieties that officers experience when confronting their transition to a civilian job are shared by almost everyone changing careers, whether uniformed or not. The most successful person at dealing with the stress is one who looks to the future, establishes a "get out" time period or date and starts preparing for the inevitable early-on in the process. At a minimum, a year is required to establish an executable plan and then carry it out. In most cases, you can't save extra money for unforeseen transition expenses, do the extensive career research required, and network with the "right" people in less than a year. The more planning and time you and your family put into preparing for your transition, the less anxiety you will all experience when that uniform is hung up for the last time. Finally, if for some reason you failed to prepare early enough or were surprised by higher headquarters with an involuntary release notice, you must hit the ground running. Do not, however, express panic in your voice, resumé, or during an interview. Instead, dash for your local transition office and ask the staff for assistance

Early Preparation

For those of you thinking about a career change or even for those of you who are happy while still on active duty, some of the following thoughts may be worth considering in preparation for leaving the military. They may also serve as an early wake up call regarding problems that could effect your job or your family down the road.

1. Check Your Attitude. Have you become complacent, lost your creativity, or stopped caring about your current job? How are you getting along with your co-workers? Ask your family and friends what they think. A bad or indifferent attitude could signal that things are not going well around the office or at home.

2. Step on the Scales. Have you put on a few pounds during the past few years? If you have gained excessive weight, resolve to start exercising regularly along with eating sensibly. Make daily exercise as much a part of your life as eating. Exercise can improve your attitude as well as your health.

3. Take a Look at Your Wardrobe. When was the last time you threw out old clothes? Even if you think no one cares how you dress, trust me they do, and possibly out of politeness, they may keep quiet about your frayed trousers, out-dated clothes, or shoes that need to be polished. Go through your entire closet and get rid of the items you haven't worn in a few years. Check what's left for serviceability, style, and fit.

4. Tune up Your Social Skills. Have you been sociable lately? Have you isolated yourself from family, friends, and coworkers? When was the last time you invited friends over or went to someone else's home (other than a relative's) for a purely social visit? Resolve to join a social or professional organization or get involved in community activities. They can provide valuable social and professional networks.

5. Read More. What was the last professional magazine, journal, or book you read? Are you watching business trends to see if your industry is changing? If so, are you changing with it? Avoid getting into the habit of slipping into a nightly coma watching TV after a few drinks and a big dinner. I've never heard of a person getting a job while watching television.

6. Take Control of Your Credit Card Balances. If you have more than a $1,000 balance on your credit card, realistically determine if you will be able to pay it off when the next bill comes. If you cannot pay it off, you may need to regain control of your finances. For starters, quit buying things you don't absolutely need until your credit card debt is zero. Also, talk with your spouse about having one credit card each and paying the full balance each month.

7. Create or Update Your Resumé. If you are still on active duty write a resumé and get it over to the base transition office for review. Do not pay anyone to write your resumé for you. If you have been in the civilian workforce for a while, review the one you originally wrote for your current job and update it.

8. Make a Long Term Financial Plan. Instead of going from month to month making ends meet, look five to 10 years out and decide where you want to be financially and professionally. Then determine what you need to accomplish your goals.

If after reviewing the above you are truly happy with your personal and professional life, stay on course.

Blind Loyalty

In the uniform services, loyalty is an absolute. Loyalty to your commander, your command, your service, and your country is never questioned and has been instilled in your culture from the first day you put on your uniform. However, as you prepare to leave the uniformed services for a civilian career, loyalty can become one of a number of challenges you'll face.

You cannot wait until the last day of service to begin preparing for a job hunt in today's highly competitive market. As you make the decision to retire, you need to direct some of your loyalty toward yourself and your family. These are often neglected during a career transition. You will also need to devote more time, resources, and attention to your job search and your retirement goals.

Marketing yourself for a civilian career is another big challenge. Unlike loyalty, marketing yourself is not part of the culture of the uniformed services. Not so in the corporate world, where marketing yourself is a technique that must be mastered if you expect to compete with your corporate contemporaries. This challenge will require considerable homework, for only a thin line separates being supremely confident in selling yourself to a potential employer and coming off as an overbearing egomaniac. Work at being humble, honest, and confident, but also work at being able to sell your many skills.

Most military personnel have business sense but can't articulate it very well. Anytime you can execute requirements within a given time frame, with available resources, and not lose personnel or resources, you have a successful business proposition. Further, you have to plan the requirement, communicate the plan through the corporate headquarters, and then get approval from the corporate executive officer before executing the "business proposal." Anyway you look at it, you have conducted business but used a different vernacular. Your challenge is to conduct research, study, read, and learn the new language for what you have been doing your entire military career.

If you do not have a computer at home or a good working knowledge of popular word processing programs, spreadsheets, and graphics, you face a huge obstacle. Computers are a necessity for information gathering, preparing resumés, and other correspondence in the job search, and in communicating with most other areas of the business community. These electronic marvels allow you to e-mail your resumé, surf the job boards, explore business opportunities and enter "communication rooms" with others who share similar career interest and goals. Computers are also a fixture in any place you'll work. Automobile mechanics use more computers than most administrative personnel.

Yet another challenge, networking, is the key to a successful job search—85 percent of all jobs are found this way. You must get out, shake hands, meet people, get involved in organizations, and ask for help in finding a job. Take this challenge and make it work for you.

Finally, deciding what you are qualified to do will probably be your greatest challenge. You do not have to pay someone to test you or to tell you what interests you most. Simply sit down and list those things you enjoy most, what you have done successfully, what you will do or not do if you have to. Apply this short exercise to any number of potential career opportunities, and think in terms of managing resources, meeting deadlines, accomplishing tasks as the "team leader;" things you have been doing all of your military career. Throw in a lot of hard work and enthusiasm and you will succeed in your transition to civilian life.

Many families spend more time discussing the purchase of a home computer than they do talking about one of the most traumatic events of their lives: the transition to the civilian community. From the first moment you think about transitioning, consider sharing that decision with every member of your family and anyone else who will be affected by the decision. Obviously the most important family member to tell is your spouse, for your spouse will be closest to you in every phase of this long, sometimes arduous journey from uniform to civilian attire. Children should be included in this decision too, for their worlds can be dramatically effected by the process.

Consider scheduling time for the entire family to sit down and talk, a time without distractions. If you broach the subject by saying, "Honey, I need your advice on what you think I'm best suited to do upon retiring next year," you'll succeed in getting the television volume turned down but maybe not much more. If you say, "I may not make as much money as I do now, and you may have to go to work," you'll clearly get your spouse's attention. And if you tell your son you may have to sell his old car because it needs a lot of expensive work and you can't afford the insurance premiums until you get another job, his CD-playing headphones will come off in a flash.

Such statements are not just attention getters, they are truer than you know, and your family needs to realize this. In my one-on-one counseling sessions, many transitioning military officers say their families don't understand their ordeal. "My wife is really out of sorts and down on me all the time, and my kids just don't understand why they can't attend expensive out-of-state college," I'm told. If you don't include the entire family in the process early on, they will not only be unhappy, but, more importantly, they won't understand the difficulties or the triumphs of the process.

Early in the process of transitioning when you're thinking about what you want to do, where you want to do it, and when you want to do it, who better to bounce an idea off of than your wife or husband—the person who knows you best and has a vested interest in the final outcome. It is not unusual to have a former military member tell me that they have not yet discussed their choice of a civilian career with anyone, only to find that their family is not happy at all about the location of their new home, timing, or financial sacrifices associated with the entire transition.

Spouses who have not worked in many years, have lived in military communities with the security of military housing—in some cases, big housing—often have no idea of the changes in life and lifestyle they are about to undertake upon their spouse's separation from the military. Like many unknowns or unpleasant events, some spouses simply do not want to get involved in the transition business and want to maintain the status quo as long as possible. But it is extremely important to encourage your spouse to attend transition briefings and job fairs, to read transition articles, and become involved in your (their) future. It can be tough finding a suitable location that maintains the standard of living your family has come accustomed to during the last 20 to 30 years. The time to talk about moving to Slippery Road, West Virginia for a civilian job should take place long before a job interview. While a family may not like some of the events in the transition, they can at least understand them if you're open.

If everyone is included in the transition plan, they will understand why you are depressed from a recent job rejection, or worse yet, no interview at all. They will understand why you're all of a sudden buying yourself new clothes and they get none. They will understand why they are eating at home more often. They will understand that times are tough.

Leaving the security of military life can be one of the most traumatic events of your life. But think of it as a team effort, for you and your family will either sink or swim together. The family should become a buoy, not an anchor, and they will if you keep them involved in the process.

Separation from the military offers a great opportunity to stop, take stock of your life, and take charge of your destiny. Most of us spend the first third of our lives in school and the second third in the military. The upcoming third phase of life will, or should, allow us—financially and emotionally—the time to fulfill some of our Walter Mitty dreams. Taking charge of your life after military retirement means much more than simply finding a job, although this is a critical piece in the puzzle of sustaining life. The transition from military to civilian life requires you to get a clear, realistic grasp of your pros, assess your financial responsibilities, and talk to your family to determine your needs and the needs of your family.

If you are honest with yourself, you'll know within three to four years before retirement exactly how your career is going and whether or not you are still in the race for advancement to the next grade. Did you make the last promotion list? Were you selected for the right staff position or command? Was your most recent evaluation report strong? These are questions you should know the answer to as you decide how much longer you will be in uniform.

Midway through your military career, or earlier if possible, you should have set some realistic financial and investment goals that would help get your children through college and still leave you a little for investments. However, if you are like most of us, you probably saved very little and spent a lot acquiring "things"— not necessarily what you needed but what you wanted. All is not lost, however. It is never too late to plan for the future.

At least three years before you intend to leave the service, plan to save, invest, or acquire enough money to get the children out on their own, care for your aging parents (if necessary), and reduce major debts. Start tightening your family's belt by only buying the things you need, not want. If you don't control your spending now, it will be hard to focus on what's important at retirement time, when nothing but money is driving your train.

Many of you will need to take a full-time job for a few years after military retirement. If so, make that second career part of your plan. Always ask yourself these questions: How much money do I really need? Is it important to leave my children a substantial inheritance? Do I need a modest condo, or a house with room for all the grandkids?

This is also the time to sit down with your entire family and plan the rest of your lives together. The average active and productive human life span is around 75 years, but as one member wrote, "You only are guaranteed one day at a time." Talk with your family about how and where to spend those days. Are there dreams that your family members put on hold during your military career? This may be the time to let their aspirations determine the direction of your shared lives.

Finally, let me highlight three key subjects:

- *First, plan to stay physically fit.* Your health is more important than all of the money in the world. Put daily and weekly exercise right up there with financial planning; without good health, nothing else matters.
- *Second, if you have let them lapse, reestablish relationships with your immediate and extended family.* Military life can take its toll on those who have stood by you during your military career.
- *Third, stay intellectually challenged.* Read books, attend plays and concerts, and get out and socialize. Make friends through hobbies, volunteering, and activities that keep you active.

Advance planning on the part of the whole family, coupled with a realistic assessment of your personal and financial needs and obligations, can make the third phase of your life the most fulfilling. Plan now to take charge of your life!

3

The "Operations" Plan

When Can You Start? Planning the Day You Go to Work

Officers planning their transition program frequently ask me the following question: "If an employer is ready to hire me, when do I tell him or her I can come to work?" The answer may appear obvious, but there is more to it than, "I'll be there when you want me." Many of you have service commitments from previous school or training assignments, a recent permanent change of station, or a two year "lock in" for accepting a promotion. Depending on your particular career specialty, your retirement date may depend on the needs of your service, regardless of your other commitments.

Once you have cleared these obstacles, you also have to consider family commitments, particularly if your new job requires relocation. Children who are in their last year of high school, a lease on a rented home, a spouse's job and retirement plans, or aging family members who depend upon you for emotional and financial assistance all must be factored into the equation. I often remind officers that a military to civilian career transition is a full time job and not something to put off until the last minute. Start to address these concerns at least one year before your anticipated retirement date or even earlier, if possible.

As you begin to prepare your transition plan and try to determine the exact date you will be available for a prospective employer, your visual plan might look like a dart board with large concentric circles with your retirement date being the bull's eye. The outer circles represent time set aside for activities like long-range estate planning, additional education, learning about the business community, and researching where you might want to live. Circles closer to the bull's eye designate preparing a networking resumé, expanding your network of friends and business associates, saving money for transition expenses, and doing more research. The smallest circles include attending job fairs, conducting informational interviews, and doing even more focused research on potential jobs. Very close to the bull's eye is a job interview and a focused resumé. The actual bull's eye is the day you plan on wearing your uniform for the last time.

Although you may have made time for all of the above, you still have to determine the exact date you can go to work at your new job. If you have planned well, you will have 60 days accrued leave, 20 days career transition time (to look for jobs and go on interviews), and 10 days to do all the paperwork, get a retirement physical, attend the available transition programs, and clear the base facilities. Do not waste these days; make every minute count because you will need them all. You have paid your dues over your many years of dedicated service, and you are entitled to use every one of your leave days.

Generally speaking, employers hire between three and six weeks out, and they are hiring on their time lines and their needs, not yours. Ideally, you will retire one week, start your job the following week, and cash in your accrued leave the next. You can do this, but you must use your time wisely. Decide the exact date you want to retire, make sure your assignment folks have approved it, and count back from that date 80 days (60 days leave plus 20 days authorized transition time). That's the earliest date you can start your new job. Don't tell an employer you are available for work and then find that you can't be there; if that happens, you've probably lost that job.

Surprisingly, many civilian employers conduct the interview, tell you that you have a job, and then take up to three months to make the final decision about your employment. The military propensity to make a decision and execute it immediately does not apply in the corporate world. It's important for you to have a leave buffer in the event a potential employer takes longer than you expect to make a final decision.

It's good planning to have a definite retirement date and to work toward that date, but if unexpected opportunities present themselves, don't hesitate to change your bull's eye date and adapt your plans accordingly. The key is to combine preparation and flexibility throughout the career transition process. Don't go without your full income for a few months because of your own poor planning!

Research is Critical

From the moment you decide to leave the military until you get your first paycheck as a civilian, you'll be engaged in a critical effort; research. In fact, because of the nature of today's ever-changing corporate work environment, you may never stop researching. It could take several job moves before you are satisfied with your position, working conditions, pay and benefits, and compatibility with colleagues.

Before starting an extensive research program, you must know where you are headed. You don't have to know exactly what you want to do after leaving military service because that is part of a larger research effort. However, you need to have a general direction in mind or you will end up wasting time researching the world in hopes of finding the perfect second career. Perfection is rarely achieved.

Research is important, but it must not paralyze you and consume all of your energies. Some people allow research to become an escape mechanism, using the process to procrastinate. But somewhere along the road a decision has to be made about which road to take. The sooner your decision is made, the easier it will be to do meaningful research and networking.

When should you begin research? There is no textbook answer to this question. Research can be done anytime in your career as you see opportunities to develop new interests. However, as a minimum, your research should begin one or two years before your expected separation from military service. This timing could be extremely critical if, for example, you decide that you need to update or upgrade your educational skills, need a different degree, or want to develop technical skills for a new field you want to enter. Degree programs not only require personal and possibly professional time but also cost money that you may not have planned for in the family budget.

Where should you start your research? This may surprise you, but start with you and your family because all of you have a vested interest in the outcome. Discuss the following questions: Are you and your family happy with what you have been doing for many years? Are you considering other careers? Do you like the area in which you live? What college plans do you and your children have? Such subjects need to be addressed before you begin researching the job market.

You'll be doing a lot of soul searching at this time as you try to balance personal happiness with financial security in your second career plans. An objective review of where you are is key to knowing where you want to go. Make a list of all the things you'd want to do if you could fulfill lifelong dreams. This job wish list will be based on your interests, your experience in previous jobs, and your hobbies and skills. Once you have established one or more fields you'd like to enter, you can start collecting information about these fields. Go to local libraries, chambers of commerce, and businesses in your field of interest; check out the job resources on the Internet, attend job fairs, and generally gather as much information as you can. (The Military Officers Association of America (MOAA) publication *Marketing Yourself for a Second Career* provides a starter list of job search publications.)

Almost all large corporations now have a home page on the World Wide Web with general information about their organizations and an electronic mail address for requesting additional information or for posting your own resumé. After gathering and reviewing this general information, you can begin to target job opportunities and specific individuals.

Gathering information for a second career is like preparing a research paper for a college course. After gathering the information, you will have to do something with it. In the case of job research, you will then lay out a plan with milestones and adhere to the plan as though a flag officer had given you a mission to complete. If you are as dedicated and attentive to detail in your career research as you were on active duty, you will be successful!

Finally, there are no shortcuts in this business. Do not be tempted to hire a job search consulting firm to do what you need to do yourself. If you cannot do this task, you are not ready to compete in the civilian work environment.

The Two-Year Planning Cycle

24 Months Before Retirement:

- Identify what careers interest you; start researching and gathering information about these fields.
- Write a transition plan detailing as much as you can about your transition to civilian status.
- Subscribe to professional journals; actively participate in related professional associations.
- Conduct informational interviews with those in your chosen field.
- If an additional certification or degree is needed, research programs and enroll.
- Enhance your computer skills; Microsoft Office and PowerPoint are a good start.
- Read books on careers and transition planning.
- Attend local base transition or other free programs.
- Observe your chosen field's dress standards; if necessary, buy a suit.
- Begin attending conferences and job fairs; get a feel for opportunities and salary ranges.
- Develop long-term network plans and never stop networking during the transition process.
- Begin accruing leave to have 60 days before last year of active duty.
- Pay off outstanding debt.

12 Months Before Retirement:

- Decide what month you will retire and inform everyone in your chain of command of the date.
  - Track your family's living expenses, and if you can't live on your retired pay for four or five months, reduce those expenses.
  - Talk with military related and similar associations about insurance and health care options.
  - Get involved in social activities.

10 Months Before Retirement:

- Check other associations about insurance and health care options.
- Get involved in social activities newspapers, professional journals, and trade publications.
- Narrow your career-field focus.
- Continue informational interviews and research.
- Draft working copies of networking resumés; send them to friends and associates for comment.

Six Months Before Retirement:

- Develop and contact a list of references.
- Stay involved with other people.

Five Months Before Retirement:
- Pay off your credit cards, and cancel all but one.
- Continue work on career-field resumés.
- Follow up with job-fair contacts.
- Continue social and professional affiliations; leave resumés.

Four Months Before Retirement:
- Accelerate your job-search efforts.
- Continue follow-up with company representatives or other contacts.
- Send targeted resumés to companies and individuals.
- Begin daily and weekly reviews of employment Web sites; place your resumé on them.
- Respond to ads in newspapers and professional journals.
- Contact headhunters (their services should be free).

Three Months Before Retirement:
- Continue your job search by responding to opportunities.
- Put a full-court press into your job-search efforts.

Two Months Before Retirement:
- Go on terminal leave.
  - Continue aggressive job hunting.
  - Finalize your job search by negotiating and accepting a position.

One Month Before Retirement:
- If not employed, continue search.
- Make networking a first priority.
- At retirement: begin your new job or continue aggressive networking.
- If not employed, continue your job search, living on retirement pay.

The Five-Paragraph Operations Order

Without a plan for your career transition, you are like a moving target hoping to get hit by a job. If you find yourself at a loss about where to start, try putting some of your service-learned organizational skills to work. One great place to begin is the simple five-paragraph **Operations Order** (OPORD). Designed for military operations, this method of organizing your time, family, skills, interest, and resources to attack your objective should get your second-career search on track.

Situation

This first paragraph requires a thorough analysis of your current military and family situation. Are you sure you're doing the right thing by leaving the military? This is the first critical step in deciding the what, where, when, and why of your next career. Make sure you understand the challenges of changing careers and are not leaving military service because of one bad boss or one bad assignment—the civilian world can be just as bad, sometimes worse.

When you write this section, include two crucial subparagraphs: the "enemy" and "friendly" situations. Analyze the potential job market as though it were the enemy situation in an OPORD. Where are the jobs? What do they pay? What type of education and skills are required to obtain, and be successful in, them? What organizations could you join to gather more information and also network with them. Who do you know in the business? These are just few of the questions to consider.

The friendly situation considers your particular skills in relation to the career held and job market you have decided to pursue. You may not be a perfect match, but you should be in the ballpark if you want to play the game successfully. Don't use your efficiency reports as a yardstick of how good you are; be humble and honest in your self-analysis.

Mission Statement

This is the most important statement you will ever write. Until you know what you want to do, you cannot organize your resources and "attack" the job market. Some simple mission statements include: seeking a position as a legal administrator; seeking a program/project management position in logistics; seeking a position in human resource training management. If you keep your mission statement short, potential employers will know exactly what you want to do. You may come up with more than one mission statement, which is fine, as long as you can execute each of them. A mission statement can also help you focus your resumé.

Execution

This section describes how you and your family will move from the military community to a civilian job with as few bumps as possible. Some steps to execute: decide exactly when you are leaving the military and what it is you want to do; conduct research about the career field; attend base transition programs; go to job fairs; conduct informational interviews; and pursue any education or certifications you'll need.

Service and Support

Here, you will lay out the services and support you will require to successfully execute your transition. What type of clothes will you need? What type of home or business equipment (an additional phone line, upgraded computer or fax machine, typing paper, reference books or CDs) will you need? Do you have your bills under control? Do you know what your living expenses are and what salary you require to maintain the standard of living you enjoyed on active duty?

Command and Signal

This paragraph contains your aggressive plan for networking with everyone you have ever known who can help you find a job. List every person or organization on the planet that can help you in your job search and make a concerted effort to contact each one for assistance.

Your completed OPORD should give direction to your career transition while allowing flexibility and enthusiasm toward any opportunity. Now, start writing.

4

Education: A Case Study

I receive many questions about the value of seeking a second or third degree to be a more competitive job applicant in the corporate world. I always recommend getting the degree but also getting some experience in the field of study. One of the first things that is important in pursuing additional education is to ask yourself, "Does it support what I want to do in the civilian world?"

This is a case study of an officer who successfully accomplished a 180-degree turn in his life and found a new career. He lamented that, "Two years ago, I couldn't give my resumé away. Twenty-seven years of a successful military career summarized on two pages of paper resulted in no calls, no replies, and no interest. I was just another highly experienced manager offering my services in a job market swamped by corporate downsizing and military force reductions. Today, companies are contacting me to ask for my resumé." Why the change?

The problem this officer faced is a common one. He had no clear idea of what he wanted to do for the rest of his life. His military career provided him with a variety of operational and managerial experiences, yet he didn't seem to have the necessary qualifications for the positions in the employment classifieds. He turned to several popular books offering advice on making a career change and enrolled in the base Transition Assistance Program (TAP).

The results were not initially reassuring. Self-assessment and long range planning take time, effort, and patience. Fortunately, during TAP, he was advised of an important Department of Veterans Affairs (VA) benefit that provides testing and counseling to help veterans understand their educational and vocational strengths and plan for appropriate educational and employment goals. (For information about this benefit call the local or regional VA office). After taking a multitude of aptitude, preference, and personality tests at a local university career services center, he received the most valuable career counseling he ever had.

His lack of career focus was, as suspected, the result of a conflict between experience, aptitudes, and aspirations. Tests indicating that he was predominantly interested in high technology careers were reassuring since he knew there were good jobs available in that particular career field. These tests also confirmed that his academic aptitude and personality type were consistent with his occupational interest. This was all wonderful except for one not so minor problem; his work experience was not in information technology. How could he overcome this skills deficit?

The answer was to start with a fresh degree. But he was already concerned about his age, and two more years in school were not going to help that perceived problem. Could the potential benefits of a new career offset the time and effort it would take to go back to school?

In this case, further research into the computer and information technology fields gave him all the inspiration he needed. Reuters cited a study that found 190,000 vacant information technology jobs nationally, and *Computerworld* estimates there are currently 200,000 such jobs unfilled. The average starting salary with a master's degree is $45,000, and salaries in the $60,000 to $90,000 range are common for managerial positions.

This information eased his anxieties about a late start. Age would be less of a negative factor in an industry short of talent and his managerial experience could be a valuable competitive strength. He still had VA benefits for educational assistance under the Montgomery GI Bill. It was just a matter of finding a university with an appropriate program, which he could qualify for and deciding whether he was willing to make the sacrifice.

The first part was relatively easy. There are many universities offering computer science and information systems (IS) degrees. There is also a new generation of IS programs based on a combination of computer science and business administration. Many of these programs are offered in the evening, and a number of them are taught on military installations. The Graduate Records Exam (GRE) may be waived if you have a high grade point average from a previous master's degree and you also may be able to transfer relevant credits from an earlier degree.

The second part was more challenging. Is it tough to go back to school as you approach the end of a military career? While individual circumstances vary, the answer is yes. A graduate degree requires a significant time commitment and you should think carefully before beginning a program. Working the long hours expected of a senior officer while attending classes in the evenings, preparing papers and presentations, and studying for exams can be difficult at times.

He gave up nearly everything but work and school for two years. Because he lacked a previous degree in computer science, he often needed additional texts to cover subjects classmates had already learned at the undergraduate level. When reading books wasn't enough, he found a tutor. The upside was that he rarely found the extra studying unpleasant because the subjects were intriguing and right in line with what he would need for his next career. It wasn't easy, but it was manageable.

The bottom line is that he received a job offer that was just what he had been aiming for. He not only had received another degree; he had a new lease on his future. It's an alternative well worth your consideration.

Developing skills needed in today's commercial economy can enhance your job security and your income potential. Taking an active approach to career transition has benefits well worth the effort. It puts you in control of your destiny instead of letting the transition control you.

5

Preparation of a Resumé

T he first words out of most officers' mouths after "I'm retiring" are "I've got to get a resumé." For some reason, a resumé seems the biggest hurdle officers confront. Yet writing one is probably the easiest, least complicated part of the career transition process. A resumé, however, will not guarantee you a job. At best, one can land you an interview, but never a job. Resumé writing is only one small part of your total strategy to successfully transition from military uniform to civilian clothes. That does not mean, of course, that you can be without a resumé. Moreover, the very process of writing one can help you sharpen your career goals. Do not pay anyone or any company to write your resumé; nor do you need to buy a computer software "resumé writer" program. You need to personally draft a resumé so you can analyze what you have done, what you liked doing, and the skills you have acquired in your military career. If you can't draft your own resumé, you have an uphill battle ahead of you to suceed in the civilian world. Writing a resumé is much like writing a paper for a college course. Gather all your reference material, like fitness evaluations, examples of resumés from people who are in or have been in the same general field as you, job descriptions of careers you may want to consider, and any other relevant references. Lay your material on a table and start writing. Obviously, some things you have done in military life will have no relevance in the civilian world. Combat awards and meritorious medals, while important to you, have very little value for an employer seeking to fill his company's needs. "Fluff" statements from fitness reports such as "best in the fleet," "rare advanced promotions," "great leader," don't tell an employer very much. Instead, describe your significant responsibilities and the results of your work if applicable to the position you seek. After every statement you make in your resumé ask yourself, "So what?" If the statement does not answer an employer's needs, it's a wasted statement.

Your first resumé will be like your first draft of a thesis, very rough and requiring a lot more thought. Initially, you should list all your major responsibilities and accomplishments chronologically in reverse order. This will become your basic, generic document from which you can "cut and paste" to craft more specific resumés targeted to specific job openings. Resumé writing is not creative writing, so don't hesitate to paraphrase from examples as you put your military responsibilities into words that civilians can understand. The perfect resumé doesn't exist. It's a living document that changes like a chameleon to fit each career opportunity. Continue fine-tuning your chronological resumé until you are comfortable that anyone reading it will understand what you have done in each of your major assignments. Then decide which areas of expertise you want to pursue and start writing a functional resumé citing responsibilities, experiences, and accomplishments in specific functional areas, such as personnel, maintenance program manager, and communications. Again, this will be a generic document until you know the position you are applying for and its requirements. Finally, work on a combination resumé that is organized by specific areas but includes a short chronological section showing a progression toward greater and greater responsibilities.

Do not attempt to write a resumé that indicates you can do anything and expect a busy employer to read it and decide what you can do best. Be clear, specific, and to the point in describing your skills, and apply them to the employer's needs. The time to discuss your flexibility, hobbies, and family is during the interview, if appropriate.

The Parts of a Resumé

At its best, a resumé might get you an interview. It is just one part of your total campaign to transition successfully from a military uniform to civilian clothes.

When you're ready to start your resumé I offer this counsel once again—Do not pay anyone to write your resumé. Don't even buy a resumé-writing software program. It's important to write your own resumé so you can analyze what you've done, what you've enjoyed, and what skills you've acquired during your military career. It will be much easier for you to explain your skills to an employer if you're the one who put them on paper.

Writing a resumé begins much like writing a paper for one of your college courses. First, gather all of your reference material, such as examples of resumés from people who have been in a similar field, job descriptions of careers you may want to consider, fitness evaluations, and books on how to write resumés. There are few original thoughts in resumé writing so don't hesitate to paraphrase these examples.

The first draft of your resumé will be very rough. List your major responsibilities and accomplishments chronologically from the most recent to the earliest. This will become your base document. Continue fine-tuning this section of your resumé until you are comfortable that anyone reading it will understand what you have done in your major assignments. Now, decide which areas of expertise you want to pursue, and start writing the functional part of your resumé, which focuses on a particular career field, like personnel, maintenance, communications, etcetera. This will be a generic section until you know exactly what job you are applying for.

Finally, turn to your specific functional skills. If you assist the employer by providing clear applications of your skills, he or she will appreciate it and may reward you with an interview.

Once you are comfortable with the content of your resumé, you can think about format. I like a simple, four-part format with a generic objective statement (i.e., "position as a financial manager of a nonprofit trade association") that you can make more specific once a position is announced. Follow this with a career summary paragraph showing how your specific skills relate to an employer's specific needs. Education follows, with degree programs listed in reverse order. (Depending on how senior you are, you might leave off dates.) Finally, list your professional experience and accomplishments. If you are pursuing a faculty position at a prestigious school or medical service college, you also may need a curriculum vitae (a short account of your career and qualifications) in addition to your resumé.

Remember, this is not hard work. It just takes focus. If you spend more than a couple days doing this, you are wasting time. Now, get moving!

Resumé Mistakes

A great resumé won't get you a job—that's what the interview is for—but a poorly written one can cost you an interview. Here are some mistakes I've seen over and over again on resumés.

Mistake No. 1: No Clear Objective

Many people don't put an objective on their resumé because they aren't sure what they want to do. If you don't know, an employer isn't going to take the time to figure it out. Employers have specific positions to fill and by clearly stating your objective you help them know where you'll fit. The objective must be the most focused statement on your resumé. It also should be realistic. You didn't go into the service as a colonel or captain, so don't expect to go into a new organization as the CEO. If you have a job description, your objective should include or restate the position title. "Seeking a position in marketing and public relations" is a great one-line objective statement for a business development opportunity.

Mistake No. 2: Lack of a Clear Summary Statement

Employers are busy. Your resumé will probably get a 10 second look before they decide to keep it or pitch it in the circular file. That's why a succinct summary statement is so important. Tell an employer what you can do for his or her organization. You must know the bottom line and how you can make an impact.

A summary statement of six to eight sentences highlighting your skills will get someone to read further. "Twenty-plus years in all facets of logistics operations; extensive hands-on field operating experience, with staff and supervisory experience throughout the Department of the Army. Most recently served as the Director of Logistics Operations at Fort Lee, Va., followed by a senior-level staff position within the Department of Defense and as an instructor in the Defense Systems Management College, Fort Belvoir, Va.," is a great summary statement.

Mistake No. 3: Failure to List Appropriate Education

In many career fields, it's more important to have current certification than a liberal arts degree that's 10 years old. Find out what professional certifications are available for your chosen field. If they are a prerequisite for employment, get them and include them at the top of your resumé. You may not get past the 10-second look without them.

Mistake No. 4: Focusing on Jobs Held Instead of Results

Many resumés list positions and performance objectives but not accomplishments. An employer wants to know how well you executed your responsibilities. Present your most significant positions and accomplishments as they relate to the job you're seeking, focusing on the last five to eight years so your skills don't seem dated. "Responsible for the staff of a large corporate headquarters and tasked to reduce its structure. Within 60 days, developed and implemented plans that reduced the staff by 30 percent while simultaneously relocating to another geographical region of the country," would capture a human resource director's attention. Another example might be, "Assumed responsibility for the maintenance operations of a large vehicle fleet and reduced maintenance backlog by 50 percent in the first quarter. Increased overall operational readiness rate to 95 percent for the entire year."

Mistake No. 5: Poor Presentation

A cardinal sin in resumé writing is misspelling a word. Poorly constructed sentences and bad verb usage run a close second. Write the way you talk. If you use words that are difficult for you to use in context, they will be difficult for the reader. Use your computer's spell-check but don't rely on it alone. Read, reread, and ask friends to read your resumé before you send it to anyone.

There is no perfect format for a resumé and everyone has an opinion on how to write one, so don't spend your life polishing the perfect resumé. Definitely don't pay anyone else to write your resumé. You will be the only one at the interview to defend it, so you must be the one to write it. Finally, before you submit a resumé, ask the organization's personnel office for instructions on how to transmit it.

6

The Cover Letter

I n the military community you don't have to present yourself to a potential employer. You simply "report" to duty, figure out what's being done and what your role is in the organization, and go to work with little guidance.

The corporate world is similar, except you must present your qualifications in writing before being considered for a job. This letter is referred to in the career transition business as a cover letter, approach letter, targeted letter, or broadcast letter. Regardless of what these letters of introduction are called, they all have one common purpose—to get a hiring official to call you for an interview and eventually offer you a job.

Several steps must be taken to ensure your letter of introduction is read with interest. First, and most important, is making sure the addressee's name is correct. This person should be someone who has the power to hire and fire. You can probably get this name through research or a networking source within the company. Do not mail a letter to "Dear Sir or Madam," "Director of Personnel," or "To whom it may concern." Such addresses will only ensure the letter ends up in the trash can. If you can't get a hiring official's name, you are probably wasting your time, envelope, and stamp.

Second, establish a link with the company or a person in the company in the first few sentences. This step will be key for someone to take the time to read further than the addressee's name. For example, "Bob Johnston in your company's marketing department suggested I apply for the job in training development" is a good lead in, particularly if Bob is in a key position within the company. Ideally, you would ask your source at the company to personally deliver your letter to the hiring official and speak on your behalf as someone who would make great contributions to the company. You can almost be assured of a call and possibly an interview if you have the minimum qualifications.

Third, state the specific position you want within the organization: "I am applying for the position of XYZ's director of training and believe my years of training management have prepared me for this job." Restating the job announcement and number, if appropriate, is helpful particularly if the company has a number of positions open and is expecting a lot of letters and resumés.

Fourth, state your qualifications for the position. Review the requirements of the job and summarize what experience you can bring to the company. Depending on your writing skills you can either address in detail your qualifications in the letter or attach a resumé for more information. Don't make the letter too long. Rather provide just enough information to interest the employer and whet their appetite.

A letter of introduction is difficult to write, so practice until you feel comfortable that you are addressing the employer's needs. After each sentence ask yourself, "So what?" If your statement doesn't make sense or doesn't answer the question, make changes. Remember, employers are busy and get plenty of applications so they will initially try to weed out as many as possible. I am not suggesting you supply the employer with fancy gimmicks—only simple, straightforward facts addressing the employer's stated requirements.

Finally, in your closing paragraph ask for an interview and provide available dates and times if appropriate. "I would like very much to be considered for the position of Director of Training and am available any afternoon for an interview." Provide your office, cell, and home telephone numbers, and close with, "I look forward to hearing from you," or "I will call you next Tuesday for an appointment."

Now that the letter is written, double check your spelling and grammar and make sure the letter is easy to read and makes sense. Does it address the employer's needs? Have you told them what contributions you can make? Would you hire the person based on your letter if you were the hiring manager?

If everything is as you want it, print the final letter on white bond paper. Check to make sure it looks neat and is centered with equal borders and that it's something you could proudly present. If so, you have a finished product. This letter of introduction should rarely ever be more than one page. Enclose a resumé of no more than two pages if you want to provide more detail.

As mentioned earlier, attempt to hand deliver the letter personally or have it delivered by someone in the company. Call after three or four days to see if the letter has been received and ask if there is any other information they need. Do not overdo the phone calls once you are sure your resumé is in the right hands. Use good judgment to decide if you should call in a few weeks to check on the status of the application and a possible interview. Also, keep track of when you mailed material, called, followed up, and provided or received input from a company. It is embarrassing to call the office and ask about information that has already been provided.

You have only begun the career transition process and writing the letter is the easiest thing you will do!

7

References

References are often the cornerstone to your transition program and essential to a successful job search. Unfortunately, many people wait until the last minute to decide who they will seek as references and end up failing to adequately exploit these valuable resources.

One of the first things you should do in transition planning is make a list, along with addresses, emails and phone numbers, of former employers who can speak or write about your skills. It may also be useful to make a list of people who have worked for you. Good references are often keys to job offers and very important to your networking system. Not only will they say good things about you when a potential employer calls, but if they know you well, they will also become counselors if you find yourself floundering in setting a career direction. If they have been your immediate boss or mentor, they will be invaluable in giving you suggested career options and networking contacts as well as critiquing drafts of your resume.

The definition of the word "reference" in the dictionary may be the key to who you select: "a person who is in a position to recommend ... [and provide a statement about another] person's qualifications, character, and dependability." The word "qualifications" is important in this definition. Character and dependability are nice personal traits but they won't get you hired if you aren't qualified to do the job. It is critical to use references who know your specific qualifications for a particular position and who can speak or write about your other outstanding qualities as well. Your hometown minister, high school football coach, or father-in-law are not good references for a company trying to fill a logistics position. In most cases, a potential employer is not concerned about how well you coached the soccer team or your frequency of church attendance. He or she wants to know how well you executed a program or project similar to the job being offered.

The following rules apply to those who are seeking and using references during the job search:

- *Before giving anyone's name as a reference, get his or her permission.* If he or she agrees, provide him or her with a copy of the job announcement, your resumé, and copies of fitness reports, if appropriate. This will help ensure that he or she remembers you, what you did, and how well you did it. (An inflated fitness report does not mean someone will remember you.) Don't be naive enough to think that every senior person remembers everyone he or she ever rated. You may also want to draft a reference letter for his or her signature. The person you ask to write a letter on your behalf may be very busy and not have the time, or desire, to put a lot of effort into a reference letter. Who better knows the requirements of the job than you, the person doing the research on the position?
- *Do not abuse a reference by using them on every application.* Remember, they are also busy and may not give as much attention to your fifth request as they did to your first. Every reference is critical!
- *Use the appropriate reference for the appropriate job.* If you are applying for a logistics position, use a logistics reference you worked with, not a staff college faculty member, golfing buddy, or someone you worked for as a pilot.
- *As a courtesy, keep your reference informed about your status.* If you don't get the job, they may be able to help you with other contacts.
- *Send a thank you note, regardless of the outcome of your search.* Not only because it's the proper thing to do, but also because you may need them again in the future.

In the process of selecting references, use a cross section of officers, non-commissioned officers, warrant officers, and civilians, if possible. You never know who is interviewing you or what position and grade they held while on active duty, if they served at all. In some instances an interviewer may be turned off by a list of senior officers as references. They may be reluctant to write or call the officers, or it may send a signal that you only know how to work in "high speed" organizations where everyone has secretaries, aides, and drivers. If their company is small and everyone does everything, including making coffee and loading the copier with paper, they may fear you are not a "sleeves up," get your hands dirty type of person. If appropriate, you might also use someone who has worked for you as a reference. Not only will it show confidence and humility on your part, but it will also show that you made an impression on people who worked for you.

A lot of thought must go into acquiring references. Don't fall behind on finding someone who will give a potential employer good, valuable information about you. They can be the key to a job offer.

8

The Desktop Briefing

s a job seeker, your mission is to target a career field. Many military members think first of the Department of Defense as an industry that can offer a civilian career. But employment opportunities have changed as the industry has downsized, merged, consolidated, and shifted its focus in response to the massive reduction of our armed forces and terrorist activities. Making the career search more difficult, fewer leaders in the corporate world have had military service or understand what skills a former military member can bring to their organization.

But the challenge to identify a career field is not as daunting as it might seem. The defense industry may be downsizing, but other industries are growing. Your military career can actually give you an edge if you market yourself well.

To succeed, you must study and continually monitor the ever-changing job market. Many changes in our economy can effect employment. Your research can lead you to career opportunities you might never have considered. Have you thought, for example, of the business, technical, and even recreational career opportunities presented by our country's burgeoning health care industry? Travel agencies, recreational activities, assisted living facilities, pharmaceutical companies, and health and fitness concerns represent but a few of the areas taking advantage of the boom in health care awareness. The electronic and security industry is also bursting with opportunities because of the revolution in consumer information, entertainment videos, telecommunications, and computers in the business world. Events that create and effect jobs include political upheaval in both friendly and adversarial nations, which caused changes in the national and international business community (For example, the trade deficit with Japan and China). The events of September 11, 2001 have changed the entire landscape of National Security operations and presented many new job opportunities in security operations, intelligence operations, and anything related to military weapons.

That moment you've been hoping for just occurred. You've been called for a job interview. So why are you worried?

Although promotion and command selection boards interview candidates through a series of screening processes, few officers have been personally interviewed. Similar screenings occur in the civilian business community, but they are performed by human resource personnel instead of command selection boards. Civilian screening starts with your resumé and perhaps a phone interview before the first face-to-face meeting at a personal interview.

Your preparation for an interview begins long before the actual event with your education and work experience. All the chemistry you might have with an interviewer is for naught if you are not qualified to do the job. On the other hand, if several applicants are nearly equal in qualifications, the person who passes the chemistry test is the person who probably gets the job.

Passing the chemistry test involves many tangible and intangible elements. Tangible elements include how you answered the telephone when the company called for an interview, how you dress for the interview, your education and work experience, and your knowledge about the company and how you will contribute to its goals.

Intangibles include your personality, sense of humor, and the ease with which you interact with the person or team during the interview. There is no single item that is critical to a successful interview; rather, it's a delicate balance of tangibles and intangibles. Some of these are within your control; others are matters of luck or timing. You have absolute control over the tangibles, so focus your initial energy on them.

Start with education and experience. If you don't have it, get it. If you have no experience in the field or a related one, you probably will not have a successful interview. In this situation you must be willing to start at the bottom and work your way up the corporate ladder—which is the way it usually works with successful corporate people. If you are qualified for a particular position, expect to start above entry level but don't expect to start at the top.

How you dress is another tangible. Do your homework. If you don't know what attire is appropriate, call and ask what they wear at the company, drop by and see for yourself, or ask others in similar fields of work what they suggest. Don't leave this to chance because if you go into an interview looking like Pee Wee Herman, you just flunked the chemistry test.

Knowledge of the organization is an important tangible that will help you sell yourself to a potential employer. You can get this information through library research, Internet searches, networking with friends, or by asking the company for year-end reports or related data.

Intangibles during an interview are difficult to address, but I sum them up as "be who you are." Relax and remain confident, as though you were talking with a friend about your career aspirations. Don't pretend to be someone you're not. If you don't fit on a team, it's better to know before you are hired and save you, and the organization, more traumatic times down the road.

Are You Ready for the Interview?

The ultimate goal of every job hunter is to get an interview because it is the culmination of all the hard work you have put into your transition program. All your networking, research, telephone calls, and resumé writing come together when you get an interview. You're excited, your adrenaline is high, your anxieties are even higher, and you don't want to blow it. Relax. It's not a big deal, if you are prepared. Just as you would train before running a marathon, you should do your homework before an interview.

Begin your homework by writing down all the questions you would ask if you were doing the interview like, "What can you tell me about yourself?" "What have been your most significant accomplishments?" "What are your biggest weaknesses?" or "How can you help this company?" Use a notebook and prepare responses to your questions. You might even have a friend or spouse do a mock interview.

If you are thorough, there won't be many questions you have not considered. No matter what the interviewer asks, you will have some idea what to say. You also may want to formulate responses to personal questions, since many interviewers are not familiar with federal regulations restricting these questions. If interviewers ask how many children you have, your marital status, your age, or similar personal questions, how will you respond? If you remind them it is an illegal question and none of their business—you will be right—but it will not help you get the job. If you can lightheartedly evade the question, an employer really wants to know only one thing: Can you contribute to their organization? Focus on what you can do for the company, and nothing else should matter, regardless of your age, sex, religion, skin color, or family size.

Now that you have done your homework and prepared for the interview, think about what you should wear, how long it will take to get to the interview location, where you will park, and any additional information you should take with you. For more confidence, remember that part of the interview already has taken place when the employer reviewed your resumé and called you to set up the interview. Being yourself and relaxed during the interview are of paramount importance.

When you enter the office, be polite to everyone, from the receptionist to the person who does the hiring and firing. Be aware that everyone in the office is interviewing you the minute you walk in, examining everything from what you wear to what you say. As you are introduced to the team, be sure to remember each name, give a firm handshake, and smile. Also remember that the interview is for both you and the company. They want to see if you fit into their organization and you want to see if the organization fits you. You are not at the interview for a beating. You are there to share your past achievements in words the interviewers understand, making clear what you can contribute to their company's future.

Answer interviewers' questions as fully as you can but don't talk too much. Look them in the eye, and take no more than two to three minutes to answer each question. If they need more information, they will ask for it. If you do not know an answer or cannot provide all the information, say so. Do not try to bluff. The deception will come back to haunt you when you can't perform as promised. Act as though you were sharing your answers with a friend. If you get the job, that's the type of environment you will want to work in.

Before the interview ends, make sure you have had an opportunity to highlight your strengths. You probably won't get a second chance. Sometime during the interview make sure you ask questions about the company's growth potential, performance bonuses, benefit programs, and salary. If this is a screening interview, it is just that, screening, not hiring. A hiring interview will go into greater detail and will involve salary and benefits negotiations. Either way, always be fully prepared. A screening interview can quickly become a hiring interview if you are a strong candidate.

When it is clear the interview is over, be sure to shake the employer's hand and thank him or her for his or her time. Always ask if he or she needs additional information, what's next in the process, and when you can expect to be notified of the decision. Make sure you promptly send a thank you note to the person who does the hiring. Even if you don't get that job, there may be other opportunities within the company. If you make a good impression, your interviewers may forward your resumé to other departments for consideration.

Finally, if you don't get the job, learn from the experience, so your next interview will have more strong moments and fewer weak ones.

"Chemistry"

Many professionals who screen applicants will tell you that only about 30 percent of a candidate's total qualifications are based on the technical aspects of the job. The other 70 percent is chemistry. Chemistry can make or break you during an interview but it's something you have almost absolute control over and, far too often, take for granted.

Chemistry is defined as "the science of the composition, structure, properties, and reactions of matter." When you first meet someone, there is usually a whole range of subtle, as well as overt, subliminal reactions. Because of the value they carry during your career transition, I want to elevate these subtle reactions to a more conscious level. This is not a discussion about beauty. If you had to be beautiful to find a job, most of us would never work again. Usually, it's not what you're born with that matters but how you package it.

Physical Chemistry

My wife and I were traveling recently when she asked if I knew a couple standing in the distance. I thought they looked familiar but couldn't figure out what was missing. When we saw them up close, it was not what was missing but what they had gained that surprised me. Honestly, it looked as though someone had put an air hose to their bottoms and blew them up—they looked like beach balls with ears. Use the same discipline after leaving the service that you did on active duty to maintain a healthy personal appearance. You can't change the way you're constructed, but if excess weight is a problem for you, exercise and diet can help.

Chemical Chemistry

Many employers have told me they won't hire someone who smokes because people who smoke lose work time taking smoke breaks, have clothes that smell smoky, have bad breath, and are sick more often than those who don't smoke. With some discipline (and help from patches, gum, and/or counseling), you can quit this expensive and unhealthy habit.

Another aspect of chemical chemistry, while not as unhealthy but sometimes as irritating as smoke, is excessive perfumes or deodorants.

Personal Chemistry

This is not about how to dress but about more personal aspects of appearance. At a recent job fair, one employer had a large bowl of unshelled peanuts available. Others offered candy, licorice, or chewing gum, and people were stopping and grabbing handfuls. Someone who had grabbed the peanuts came over to talk with me and I was not impressed with his peanut breath or the peanuts stuck between his teeth. An employer would not be impressed either. Before an interview, brush your teeth and gargle. If it's an interview over a meal, avoid foods that will leave you with bad breath or that get caught in your teeth. If you want to chew some gum or use a breath mint prior to a meeting do so in the privacy of your office or car but don't show up looking like a cow chewing your cud.

Hair can be a sensitive subject for some. If you are going bald, going gracefully is often preferable to a poorly fitting toupee or other attempts to conceal your scalp. If you feel you must color your hair, get it done professionally. If you wear jewelry, don't overdo it. A few simple and conservative pieces are fine.

Chemistry plays an important role in your career transition but many of us never give it a second thought. The bottom line of the career transition process (as well as life in general) is to be whoever you are and hope you are what an employer wants.

The Desktop Briefing

The desktop briefing (also called a desk-side briefing or flag brief) is given to flag or general officers in their offices, at their desks. Probably every officer has given a briefing to a senior military officer, but most may never have thought about giving a desktop briefing to a potential civilian employer. Yet many other officers have told me the desktop briefing was the one thing that separated them from their competitors during an interview and throughout the job search.

An officer hired by Ford Motor Company stated, "During my last interview, it appeared the interviewer was trying to establish my ability to orally communicate. I promptly pulled out my desktop briefing slide presentation. I was able to outline, present, and summarize my career. ... I sold my experience in a very convincing briefing. The interviewer was totally delighted! Three days later I received a telephone call asking me if I was interested in working for Ford."

The officer used a standard, three-ring, flat notebook for his presentation. Others have used a triangular fold briefing book or a large "butcher paper" format. Today, a computer generated slide presentation should be a piece of cake for most of you in the job market. But the substance of the briefing is far more important than the mode (although a little form over substance is not always bad).

Here are a few good reasons why you should go to the trouble of preparing a desktop briefing for a potential employer.

First, if you were an employer interviewing eight possible candidates for a single position and one of the candidates enthusiastically told you not only about him or her but also how his or her training and experience could effect your organization, who would you hire?

Second, even if you don't have an opportunity to present your briefing, you will be better prepared for the interview if you have gone through the process of researching, analyzing, and understanding your needs and, more importantly, those of your potential employer.

Third, regardless of how knowledgeable and polished you are, there is always a certain level of apprehension associated with the interview process. A desktop briefing allows you to control the flow of information in a systematic and organized manner. It can prevent you from rambling and reduce the number of open-ended questions that might get you in trouble if you don't have time to think through your answers.

Fourth, preparing and presenting briefings is an integral part of most jobs today. You have an opportunity to showcase this skill to an employer.

Finally, if you get called for an interview, you can assume you are qualified for the job. The only thing that separates you from the other, equally qualified candidates might be your desktop briefing.

There is no standard structure to a desktop briefing, because it will have to be tailored for each interview. However, there are always two major parts: you and your goals, and your employer and his or her goals. Regardless of how you blend these in your briefing, remember that the employer's needs will come first with your interviewer.

A former service member prepared for an interview to be director of facilities and grounds for a small town by taking the job description provided with the application and developing a six-part briefing that included a mission statement, an outline of his values, short-term (90 days), mid-term (one year), and long term (two to five years) goals, and why he was the person for the job. He jazzed the briefing up with clip art of local landmarks from the town's Web page. At the interview, he asked the five-member panel of town and school leaders if he could give his presentation before they asked questions. They agreed and had very few questions afterward. He believes his presentation caught them by surprise but also clearly showed he was prepared to take on the responsibilities of the position and had given it a lot of thought. The bottom line: he felt in control during the entire interview.

Being Well Organized—A Desktop Briefing Case Study

In my lectures I have discussed a **desk side briefing** as a means to better prepare yourself—and your interview panel—for that all important face-to-face meeting. Cmdr. Mike Weaver, USN-Ret., went one better. He prepared a briefing book and made copies for each of his interviewers. His turned out to be a success story as a result of a lot of work on his part, and it offers many worthy teaching points, regardless of the type of position you may be seeking. Here's how his story unfolded.

Weaver attended one of TOPS' "Marketing Yourself for a Second Career" lectures given at the Pentagon. Then he came to see me about his efforts to obtain a position as director of a Naval Junior ROTC program in a Florida school system. Through networking, he had found that four school systems were considering establishing such a program within their respective districts.

After going through the application process, Weaver was able to arrange his interviews so that his first interview was with the school he was least interested in, and the last was with the one he really wanted. He thought that by the time he got to the last interview, he would be warmed up and ready for any questions the panelists could ask, and he was.

Many times, an initial interview is little more than a get-to-know you affair. But, as Weaver's experience demonstrates, leaving something more substantial than a basic resumé for those who do the hiring allows them to review the information at their leisure, getting to know the prospect's ideas and action plan in much greater detail. One of Weaver's interviewers told him she had passed around his briefing book, so he reached others outside the immediate interview panel.

Weaver's briefing book had three parts: a resumé, references, and his presentation. In the presentation, he organized 10 one-page documents, such as "Background" (highlighting his 20-year Navy experience); "Why me?" (listing key reasons why he should be hired for the position); and "Naval Junior ROTC Program" (giving an overview of the proposed program to show the panelists he had done his homework and knew what he was talking about). Then he presented his "Short-term," "Mid-term," and "Long-term" goals followed by a page detailing, "How do we get there?". Because the position was in a school environment, he also discussed "Resource Management" and "Community Service." He closed with "Contributions," showing how his naval experience and personal life made him the right person for the job.

If you provide a briefing book like Weaver's, it gives them ready-made questions to ask, questions to which you already know the answers.

While Weaver's briefing book was for an ROTC program, the format could be tailored to any profession. If you were to develop a similar presentation, rehearse it as though your job depended on it, and make copies for all involved, you, too, likely would be successful in your job hunt.

Because this was a Naval Junior ROTC program interview, Weaver wore his more casual white Navy summer uniform and gave his best shot at being relaxed to avoid giving the impression he was a strict military disciplinarian. All of his planning and preparation were fruitful, as Weaver received immediate offers from three of the schools, including the one he wanted most.

Phone Interview

There are a number of reasons employers use a telephone interview; the foremost are time and cost. After receiving a large number of resumés in response to a company's job announcement, many hiring officials will select the most promising candidates and then do a preliminary screening to select those who will actually come in for a face-to-face interview. This saves time and money and narrows the candidate field to those most likely to be hired.

A telephone interview also allows an employer to find out how well you handle yourself on the phone. This is particularly important if you are being considered for a customer service position or any position requiring extensive telephone skills, important in almost any job.

Some things to keep in mind for a phone interview: Concentrate on listening more closely to the questions, because the person on the other end of the phone may not be clear or concise, and you won't be able to see facial expressions and body movements that often convey meaning better than words. Eliminate distracting noise such as the television, radio, or family conversations. Asking over and over for a question to be repeated may irritate the interviewer. If possible, move to a quiet location where you can concentrate on the interview.

Although it's not part of the actual interview, it's a good idea to consider the outgoing greeting you leave on your answering machine, which may have as much of an impact on your chances as an actual interview. If your message is silly, off-color, or unintelligible, a potential employer may eliminate you as a candidate. Finally, do you have voice mail or call waiting in case you're on the phone when a potential employer tries to call you? If you were an employer with a long list of fully qualified candidates, how many times would you try to call someone?

What else should you do to prepare for a phone interview? Pretty much exactly what you would for a face-to-face meeting. Do your homework researching the organization you are interviewing with. What is its bottom line? How can you contribute to it? Will you fit in with the team? Are there growth opportunities? What about the benefits package? These are all things you should normally be prepared to answer or ask in a face-to-face interview.

Prepare a three to four minute introduction of yourself in relation to the employer's needs and, if appropriate, an electronic slide presentation for a larger interview panel, even for a telephone interview. Invariably, the first words out of an employer's mouth are, "Tell me about yourself," with the implied, "Why should we hire you?" Whether you are on the phone or face-to-face, you should be ready to answer.

Recently, an officer prepared for a face-to-face interview with an out of state company. After getting a copy of the job description, he went through several steps to prepare for his interview including creating an electronic slide presentation. When the organization decided to conduct a telephone interview instead, he was ready. He made notes that addressed his abilities and created briefing charts on how he would perform in the position for which he was being considered. He rehearsed (and rehearsed and rehearsed) these so when the interviewer called, he'd be ready. He e-mailed charts to a company human resources person (after asking permission), verified that each panelist had access to all the charts, and answered panel questions briefly and succinctly. He made sure not to do all the talking and remained as relaxed as possible during the interview.

Because he could not give his presentation in person, he asked if he could attach his slides to an email, then had his interviewers load the slides on their computer so he could give them his planned briefing over the phone. They agreed and were impressed with his creativity and his ability to organize his thoughts. He was hired on the spot.

If you were a hiring authority and a candidate went through the above process for you, wouldn't he or she make an impression?

9

Wardrobe and Appearance

Are you planning to leave the military soon? What part of the transition process do you think is most expensive? A new computer? Travel? Long distance phone calls? Guess again. It's your new wardrobe that will put a sizable dent in your savings.

Dressing for the corporate world is probably a new experience for you. In the military, manuals showed you what to wear for every occasion. The blue blazer and khaki slacks that hang in your closet only met the minimum requirements for the few civilian functions you attended, and if more sophistication was required, the good old "leisure suit," after a quick cleaning, plus your comfortable dress military shoes could have you ready for any social occasion.

Now is the time to start saving your money and planning for your new wardrobe. Your first priority should be the "interview suit." While you may think you can run over to the military exchange and buy a suit, think again. Exchanges carry a great selection of casual clothes and sportswear but not necessarily the quality, or style, of dress clothes you need for the corporate world. No matter how smart, humorous, or good looking you are, if you are not dressed appropriately for an interview, you may have lost the opportunity to present yourself as the best candidate for the job.

Once you've bought your interview suit, it's time to think about the rest of your wardrobe. Different occupations and regions of the country have different styles and dress codes. You don't want to invest in Western clothes and find that your next job is in the Northeast, where people dress more conservatively. You don't want to buy a heavy wool suit, or dress, and end up along the Florida coast. Academics have different styles of dress than bankers and lawyers; evening dress in California is much more casual than in Virginia; and some "contractors" require a white shirt and dark suit for men and conservative dress and heels for women for every visit to the Pentagon. So save your money and wait until you know exactly what you'll be doing and where your next career will take you before you buy a new wardrobe.

How do you start your search for the right wardrobe? If you're a man you can never go wrong with your first interview suit if it's 100 percent lightweight worsted wool, dark blue or gray, classic in style, worn with a white shirt, burgundy tie, and a pair of dark, cordovan laced shoes with matching belt. The only thing that would ruin this outfit is wearing a pair of Buddy Holly horned rimmed glasses and a pair of ankle length socks that fall down around your shoes when you cross your legs.

Some of the brightest people on this planet don't know how to dress and it doesn't matter because they're not looking for work. For those moving from a military to a civilian career, it's a different story.

Clothes have no bearing on mental skill but appearance does have some bearing on whether or not an employer ever gets around to checking out your mental skills. You might have a Ph.D. sticking out of every pocket when you arrive for an interview, but if you don't look the part, you probably won't get past the receptionist's desk.

The following tips on how to purchase and take care of your wardrobe were taken from a booklet provided by Men's Wearhouse available at many nationwide retail stores. (Men's Wearhouse offers a 10 percent discount to all military personnel on their intial purchase.) Women have more flexibility when putting together their career wardrobes, but some of the following tips—such as on fabric and fit—still apply.

Your basic interview suit should be a dark color; charcoal gray, navy blue, or black in either solid or pinstripe. A brown suit is fine for your wardrobe but not for an interview. A single breasted suit is appropriate for all fields of employment. It's a classic that will never go out of style. Four-button suits may work in a creative environment but are too fashion-forward for some.

The fabric should be 100% worsted wool. Wool is a natural fiber that breathes, which means you will feel more comfortable. Look for suit jackets that are fully-lined and pants that are lined to the knee. Lining increases comfort and durability while reducing wrinkles in your garment.

Do not buy size, buy fit; your suit always should feel comfortable. The jacket collar should follow closely the silhouette of the neck, with no gaping. The jacket also should lie smoothly over your shoulders and across your back. It should be long enough to cover your entire seat and look proportional to your physique. Your jacket sleeves should fall just at or below the break of your wrist.

Most suits come with pants that are either double or triple pleated. A cuffed hem is traditional and preferable. The pants should feel fuller through the thigh and should be worn right at your waist.

Unbutton your suit coat when sitting. When you are in a car, make sure you hang the jacket up. Use curved hangers. Hang suits with the curve going forward. Leave space between garments in your closet.

Always wear a long-sleeve dress shirt for an interview or business occasion. There is no such thing as a short-sleeve dress shirt. Your tie should be businesslike and should complete a professional look; 100 percent silk ties are recommended, as they assure the best knot. Keep accouterments to a minimum. Suspenders or a well-chosen pocket silk will complement your look, but earrings and excessive jewelry detract from your professional appearance. Never wear both suspenders and a belt.

Lace-up shoes look professional and work especially well with suits. Choose either wing tip, cap toe, or split toe styles. Black or burgundy shoes can be worn with navy blue or gray suits. Brown shoes are not recommended for interviewing. Never wear a penny loafer or a casual loafer.

A white, 100% cotton, point collar shirt (no button-down collars) is the most appropriate interview shirt. Observe employees' shirts during your interview to judge the appropriateness of colors. Suits worn regularly usually need to be dry cleaned only a few times a year. If your suit is wrinkled, but not soiled, have it pressed. It will last much longer.

At a department store or specialty store, not taking advantage of sale prices, women can expect to pay about $400 for an interview suit, $100 or more for a pair of dark leather pumps, and $100 for a silk shirt. Don't forget to dress your outfit up with earrings and a necklace or pin. To fill out your wardrobe, you'll need two dresses, a pair of slacks, a couple of skirts and blouses, and an additional pair of shoes.

Do not buy cheap clothes that will have to be replaced frequently and often fit poorly. Look for quality clothes, decide what you need, and if you have time, wait for the sales that usually come after Christmas and at the end of the summer. If you can't wait for sales, then still buy quality clothes. You will feel like a million dollars, walk with more confidence, and get longer wear out of your wardrobe.

Finally, buy clothes that are comfortable (maybe a size larger than your dress military uniform). Nice fitting clothes will require minimum alterations. If they require major alterations to fit you try another brand. Also, good leather shoes should feel comfortable the minute you put them on and they should not require extensive "breaking in." So plan ahead for the expense of a new wardrobe. To make a successful transition to the corporate world, you have to "look the part."

10

All About Attitude

ttitude is everything in the business world, the sports arena, and in your personal relations. It is profoundly important in your career transition. Without a positive approach to the very difficult venture of changing careers, you might not arrive at the right place to retire or the right job. It is profoundly important in your career transition.

According to *The American Heritage Dictionary*, "Attitude is a state of mind or feeling with regard to some matter; disposition." As you begin to think about and plan for changing your career, your state of mind becomes important not only to you and your family but to your active duty coworkers as well. They'll be listening to your every word about your current assignment and about your proposed career plans because it's only a matter of time before they'll be in the same situation.

All too often, people leaving the service tend to say negative things about their respective service, current job, or immediate supervisor, particularly if they have been involuntarily separated under the services' downsizing initiatives. But those who remain in the service want to feel good about themselves and their service to our nation and they are uncomfortable when others "badmouth" what they enjoy. Potential employers do not want to hear your negative feelings about military service either, primarily because they will think you might bring a negative attitude to their organization. Most people can't help you with your career transition anxieties and possible frustrations, but even if they could, they might not be so inclined if you are constantly complaining.

If, on the other hand, you are upbeat and positive about life, others are more inclined to seek your friendship and counsel and to offer help when you need it. Just as important, you'll succeed at your tasks. If you believe you can find a job, make the necessary preparations, including networking, and then throw your full enthusiasm and energy into the effort, you will be successful.

To a prospective employer, enthusiasm can be more important than intelligence. Very few endeavors in the business world are singular actions. They are integrated into a larger system involving other people and usually *many* other people. Goals are achieved through teamwork. If you cannot lead the team, follow the team, or in some way contribute to the team effort, you probably won't be selected to be part of an organization.

An employer wants to be satisfied that you, as an employee, can do the work, will do the work, and most importantly, "will fit on the company team."

That's not to say that maintaining a positive attitude is easy. The most important, yet most difficult time, to maintain a positive attitude is after receiving your first rejection and it will happen sometime during your job search. The key to handling rejection is not to take it personally. It is, in fact, part of the job search process. Anyone looking for a job needs to accept the rejection, learn from it, and continue the job search.

A positive attitude is contagious. A final way to impress a prospective employer is through writing a thank you note. (Refer to Chapter 19 on thank you notes for more information.) After an interview, drop members of the interview team a personal note of thanks, even if you were rejected for a job. Members of an interview team will be more likely to remember you positively for your note. They might reconsider you for the job, or they may consider you for another position within their organization where you may fit better. Never underestimate the power of a bright and cheerful attitude. The quality of your future depends on it.

11

Understanding the Employer's Needs

hat do employers want from former service members? Many officers ask this question as they confront the task of finding a civilian job. Civilian employers want the same thing from former military officers that they want from any employee: help in increasing their profits or reducing their losses. This is the "bottom line" in the corporate world, where profit and loss statements have the same impact as a major command inspection report does in the military system. Flunk a command inspection in the military and you are history; lose money in the corporate world and you are also history.

The military system focuses more on adherence to rules and regulations; the corporate system focuses more on creativity and profit. This doesn't mean that officers are not creative. Once they understand the rules, officers, in fact, can play the corporate business game better than many in the game today.

Retiring or separating officers often take for granted the skills they acquired during their military careers. Leadership, integrity, team building, and communication are but a few. Either you were good at these skills or you did not survive the military system. Most important is your ability to make things happen with a minimum of guidance as a team leader or team player.

In the highly specialized corporate world, very few people have the leadership skills to put a team together and come out with a finished product or juggle a number of "glass balls" without dropping a few. But you can! Most service members never had the same job twice and each task was a new challenge. You figured out a way to accomplish whatever mission you were given. It's no different in the civilian world.

"If we are so great, why aren't we all employed?" you may ask. The answer lies in how well you market yourself. Military people are not taught to market themselves. It is considered inappropriate for officers to brag about their accomplishments. However, marketing yourself is an art that must be learned if you are to compete successfully in the business world. It is an art that many civilians have fine-tuned as they competed for jobs.

A significant part of marketing yourself is to convince others that you will work and "make the coffee." Many employers have never served in the military and, even if they have, their perspective on what senior officers do can differ vastly from the truth. The media, through characters like Beetle Bailey in the comics and "Major Dad" in the television sitcom, rarely portray an officer as hard working and dedicated.

Some civilians think because you draw retired pay that you don't need to work or will not work hard or do things you consider beneath you.

Your challenge–while networking and interviewing–is to convince others that you are capable of, and enthusiastic about, learning their corporate culture, understanding profit and loss columns, communicating verbally and in writing, and working as hard as or harder than anyone in their organization.

During your transition to the civilian sector, you must walk a thin line between projecting confidence and self-assurance and alienating others with egotism. Most of all, don't shortchange yourself when selling your skills. Confident humility requires attention and practice.

Finally, the business community expects you to be computer literate. If you don't own a computer and don't know some word processing or graphics programs, you are a few giant steps behind your peers. Buying a computer could be one of your wisest investments. You'll need it for work and also in marketing yourself for a second, or subsequent, career.

The Bottom Line

As you prepare yourself for the transition from military life to the corporate world, your research efforts will focus on resumés, interviewing, salary negotiations, and clothing. Although these things are important, there is one subject that many of us overlook, "the bottom line." In any type of commercial enterprise, "the bottom line" is profit, period. A member of the Military Officers Association of America recently explained this "profit" concept in words that even an old infantry aviator like me can understand. Here is what he said: "As a job hunter, you should consider profit and its place in business and in your employment. The press and TV announcers will refer sarcastically to the "big" profits made by a particular company. You may have heard your fellow officers condemning a vendor for making "too much" profit. Some organizations brag about being "non-profit," as if it makes that organization a better provider to its customers.

In a stockholder owned company, there is an absolute need for profit. Although it seems obvious, this absolute need for profit may be a whole new concept for you because it was absent in your military and school experiences. You should understand and appreciate profit and its important place in business and, thus, in your future. A good product, a noble service to humankind, and everything else come second to profit. There is only one reason for a company to exist, and that is to generate profit. Only then can a good product or a noble service be provided. Without profit, there is no reason for a company to hire you or to continue your employment.

Most companies on your job-hunting list issue shares of stock that are owned by individuals or funds. They are placing their capital at risk in order to make more money than they would in a bank or credit union savings account. For investors, there is no such thing as too much profit. However, low or negative profit is totally unacceptable.

The simple profit equation is:

Profit = Income from Sales, less the Cost of Sales

Income from sales, the first part of the equation, is established by selling price and volume. These are pretty much controlled by the marketplace. Costs, the second part of the equation, include materials, wages, benefits, assets being used up (depreciation), management pay, community support, and so forth. The company can control costs to varying degrees, but the best way for the company to maximize profits is to minimize costs. As an employee, you are part of those costs.

As a job hunter and a potential employee, how do you fit into this picture? As mentioned above, your wages contribute to costs. Examine the profit equation. The company pays you from the profit pool. If the value of your contribution to the company exceeds the total cost of your employment, you have made a contribution to profit and are a bargain. If not, you have not contributed, and the company would be better off without you. You must understand this vital concept as you start your job search.

The cost of employing a professional is on the order of twice the wages offered. This includes actual wages plus benefits, such as medical insurance, life insurance, retirement, vacation and holiday time, office materials, support people, recruiting, phone service, Social Security, and so forth.

How will you fit into the profit picture? Suppose you are a manufacturing engineer whose job is to reduce production costs. Based on the twice the wages cost of employment, you must produce annual savings in manufacturing costs of about twice your wages just to break even.

Consider a salesperson. The net profit for a typical company is in the range of 10 percent of gross sales, depending on the industry. The salesperson must generate sales such that 10 percent of those sales equal twice his or her wages, a multiplier of about 20 times his or her wages, just to break even.

Examine the job that you want. How does it fit into the profit equation? What is its break-even point of contribution to cost? Will your contribution be adequate to justify your employment?

When you apply for a job, be sure you can identify to yourself (and to the company) the ways in which you will have a positive impact on company profits. Know how you will fit into the profit equation. Be sure you are dedicated to controlling costs and be ready to show that the company will be better off with you than without you.

In short, have a clear, and simply presented approach to the "bottom line" during your job search. If you cannot look an employer in the eye and explain to him or her how you can increase profits, reduce costs, or in some way enhance the company's profit driven goals, you may not be hired.

12

Networking

L ooking for your first civilian job or looking for a better civilian job? Your success will depend largely on one factor: networking. Prepare to make it a lifelong pursuit.

Most transitioning officers must learn how to network, for networking is neither part of the military career mobility nor the military culture. But in the corporate world, nothing is more important than a network of friends, clubs, and business associates if you are to remain competitive for better jobs and upward mobility. The older you are the more important networking is because very few people with your education and skills get jobs from newspaper ads.

If you are transitioning from the service, make networking the target of your energies as soon as possible. Eighty five percent of all jobs are found through networking, so 85 percent of your energy should be expended in this aspect of the job search.

To some, networking conjures an image of politicking and can be perceived negatively. But it is critical to a successful career transition and must be effectively used not only in your initial career change but also because of uncertainty in the civilian employment cycles throughout your working life.

When should you start networking? In most cases, you have been doing a form of networking throughout your military career, although you probably don't think of it that way. The people you have worked for, with, and supervised are all part of a network who can attest to your "'reputation." You have very little control over it, but your professional reputation is based on your work skills, how you interact with others, and the manner in which you conduct your personal life, your moral and ethical character.

The other kind of networking, which is more active, can be controlled. You must consciously develop contacts through social and professional activities, through organizations, clubs, schools, and church. Networking creates opportunities to meet people who hire people or know people in hiring capacities. It also serves you by providing information about the personalities, salary ranges, and long range goals of an organization, for example, along with other information that could prove valuable during an interview. Networking creates opportunities to meet an organization's key players in informal situations, allows you to learn about job openings before they are advertised, and gives an employer information about you. One networking contact can lead you to another that eventually leads you to a job. If all other things were equal, would you hire a stranger or someone you know? Networking is advantageous both to job seekers and employers.

Where should you network? After you have decided what you want to do, and possibly where you want to do it, you should target a list of people and organizations in the field you have chosen. In some instances you might join a local chapter of an organization you have targeted. In other cases you will join an organization simply to meet people who know other people, such as the Kiwanis and the American Legion. Many of these chapters have designated networking points of contact; others are doing so every day. Chapter members could provide valuable information about the local job market or quality of life issues in the region and chapters are a great social outlet with people who have military backgrounds.

Like any organization you may join, get involved and let others see you make things happen. They in turn will make things happen for you. Finally, if the going gets tough, there is no better source of solace than having a good friend you met through networking.

13

Temporary Work and Volunteering

emporary work, or "temping," is probably one of the most misunderstood and least used employment opportunities available to officers transitioning to the civilian work force. If you are like many officers leaving the service, you are not sure what you want to do in your next career. Temping allows you to gain experience in fields in which you may have an interest but no experience, to "test drive" a company to see if you want to work there, to identify key figures in the organization for later networking, and to maintain financial stability while you look for the right job. According to a National Association of Temporary and Staffing Services (NATSS) survey, approximately 35 percent of temps are hired in to full time positions.

Temporary placement agencies were once primarily a resource for secretaries, data processors, manual laborers, and other entry-level workers, many of whom lacked technical skills and experience. While these positions still make up a high proportion of the placements provided by temp agencies, technical placement services have come into the labor market in recent years to fill upper level staff requirements.

Why would an employer use a temp agency to fill upper level positions? These positions frequently require special certification, appropriate degrees, long term experience, or highly technical skills beyond those required by entry level positions. With certification and degrees, employers also expect a more professional employee with management and leadership experience.

Employers may have special requirements for projects they want to test, onetime programs requiring special leadership or management expertise, or other specific needs requiring unique technical experience. Today, many in the business community hire at the lowest level of steady production and add temporary employees during peak periods of production in order to save labor costs.

During a recent visit to NATSS, I learned that many employers also use temp agencies when the cost of training employees becomes a burden. Employers use temporary hiring as a screening device before they commit themselves and their company's resources to long term pay and benefits. Employers are reluctant to hire untrained individuals, train them at great expense, and then find they have an employee who will not come to work regularly, enthusiastically, and on time, be mentally and physically fresh, and put in a full day's work—in other words—who does not have a work ethic.

Military personnel do not have this problem. The strong military work ethic encourages me to recommend that officers of any grade take almost any entry-level position just to get their foot in the door. They will probably move up in the organization simply by being loyal workers and keeping the same work habits they acquired in uniform.

Finding a temporary employment service is as easy as looking in the yellow pages of the telephone directory under "employment." You can also use the Directory of Executive Recruiters.

I have two points of caution: First, do not, I repeat, do not, pay any recruiter to find you a job. Recruiters should be paid by the hiring company. Second, call and ask what types of positions the temp agency places before you drive across town to find that it is an entry-level, administrative placement agency. Ask for the names of some of its client companies, the types of positions and their pay ranges, what training (if any) the agency or company provides, and what benefits are available before you are sent to a work site.

Some temp agencies offer basic benefits in addition to training; always ask. For senior officers who have never acquired basic office skills, or whose skills are not current, temporary employment agencies are a good place to brush up because you need those skills to be marketable. Downsizing started by eliminating secretaries and giving you their responsibilities, so don't be surprised if an agency asks you to take a typing test or do some other computer related task.

Finally, if you want to work for a particular corporation but can't seem to get past the human resources door by networking or with your resumé, ask someone in human resources what agency the company uses to hire temporary employees. With this information in hand, call the agency and apply for a temp position. During the initial interview process, tell the agency that you would like a permanent or full-time position with the company you hope to work for. Once you are hired as a temp at the company, start looking around for those with influence in the company or for others who can help you and begin networking.

Volunteering

This is not about the college football team in Tennessee called the "Volunteers." It's about another group of volunteers, military veterans. Retirees and people entering the work force who volunteer to help others are the ones who find jobs. Is this a coincidence? No! Let's look at why people get hired and how volunteering can help.

Education and experience are not necessarily the deciding factors in the hiring process. The "chemistry" between you and the hiring person is often a strong element in the equation. Everyone reading this book reads at a higher level than the national average. Most retired people never had the same job twice and received very little training when they took on a new job, yet time and again they figured out what to do and accomplished the task. From these two facts, the obvious conclusion to draw is that you can do most jobs if given the opportunity. It's chemistry that will get you that opportunity. How do you create chemistry? You already know the answer: networking.

Whenever I discuss networking as a means of finding a job, volunteering comes to my mind. Most people in the career transition/networking mode cultivate relationships horizontally with those on their professional, social, and economic levels and vertically with those in positions above and below them. Volunteer work falls in the "indirect" networking category, where purposes overlap. In this case, the primary purpose is to network and eventually find a job; the secondary purpose is the volunteer work.

If you are considering separating or retiring from your service in the next few years, there is no better way to start a personal networking program than as a volunteer. Local, military related, organizations, on-base retiree programs, and community sports programs are always seeking people to help. If you are visible as an enthusiastic volunteer, other people serving on your committee or program will establish a bond with you that transcends education and work experience. Then, when the conversation turns to your professional life, if you make some comment like "I guess I'll have to start looking for a job when I retire this spring," your colleague is more likely to respond, "Give me a call, and let me know how I can help."

During some of my travels, I met three retired officers who volunteer full time with their installations' retirement and community activities programs. Ross Strode is 80 years old and puts in a full day as part of the retiree program at Wright Patterson Air Force Base (AFB) in Ohio; his counterpart at Barksdale AFB in Louisiana, Steve dePyssler, not quite 80, seems to be in perpetual motion while involved in retiree activities. Lees Broome, also at Barksdale AFB, volunteers more than 3,000 hours a year, in addition to serving as the director of his local Air Force Association, a member of his local military organization, and a 12 year member of the Air Force Retiree Council.

By virtue of their volunteer work, each of these retirees knows more employers than anyone in town. Employers call them all the time to ask if they know any officers looking for work. They could probably take these jobs themselves, but at this stage in their lives they are more inclined to pass the information along and they're more likely to pass it to someone who volunteers with them every day than to a stranger.

Many of the employment success stories I've heard come from those who have found a job while volunteering and more than one officer has told me of a stint as a soccer or swim coach at a local school that led to a job opportunity. I've also heard many employment success stories from people who volunteered at their church. They let it be known that they were looking for work and someone told someone who told someone who offered them a job.

Many of those who are fully retired, volunteer selflessly for no other reason than to give something back to those who need their help. Others volunteer for the satisfaction of being involved in the community or to keep their minds active. For still others, the primary purpose of volunteering is to increase their odds of finding a job. The old military mentality that most of us grew up with might characterize this as self-serving and therefore distasteful and unacceptable. It is neither. Rather, it is a viable transition tool available to all job seekers and cultivated in the civilian workforce. Let there be no misunderstanding: There is absolutely nothing wrong with volunteering for the purpose of meeting others who may be able to help you find a job. Any time you are getting out, volunteering, and meeting people is better spent than sitting in front of the television.

So get up, go out, look around your community, and find an activity that can use your help. You have just taken your first big step toward being a success in the civilian job market, and you might have some fun while you're at it.

14

How to Work
Job Fairs

Many military officers avoid attending job fairs because they do not know what they want to do or because some fairs appear to be outside their areas of interest. But attending a job fair can give you new ideas and career opportunities. Opportunities can come from organizations you may least expect to relate to your background if you are willing to engage in dialogue with any potential employer (regardless of the industry).

A job fair attendee received a call from a Ford Motor Company representative five months after he attended a job fair. He prepared well for the interview and is now employed as a line supervisor in one of Ford's "stamping" departments (a position with which he had zero experience when hired). I've heard many similar stories of officers finding success through job fairs.

Some fairs have a specific career field focus, such as information technology, education, logistics, or management and business professionals. Others are open to any job title. A trucking industry job fair may offer challenging positions as maintenance managers, trainers, operations managers, financial officers, human resource managers, and so on. Be willing to talk with anyone and everyone when you visit these job fairs. Sometimes "chemistry" will outweigh what you assume are the professional requirements of a job. As a former service member, you can handle just about any civilian job, but only if you know about it and try.

Scan local newspapers and professional magazines to find out about upcoming job fairs. Once you find a fair to attend, consider it your first interview. You may quickly be out of the running for a potential job if you are not prepared or present yourself poorly. Wear appropriate attire, whatever that is for your particular industry and part of the country. You may have three doctorates, but if you dress like Pee Wee Herman, you will be eliminated in a second.

At one job fair approximately 80 firms were represented. Not one male in the recruiting booths wore an earring, had a ponytail, or needed a shave. Many were dressed "business casual" (open shirts and slacks), but none wore hiking boots or running shoes. The men and women in these booths represented their companies with a professional appearance. As an attendee, you should follow their lead.

Arrive early, take a stack of networking resumés, and have a list of references in case they are requested. Recruiters may ask one or two job specific questions to give you an opportunity to sell yourself in a 30 second answer. Then they want you to move on so they can meet with others. Talking with human resources folks at a booth can be a real challenge because you are trying to isolate them for a few minutes and sell yourself while others are standing making noise in the background. Pay attention to your instincts, and if you are wasting time with a potential employer, be polite and move on.

Don't waste time. Talk to as many potential employers as possible. If you have done your homework, move directly to the companies you are interested in but don't miss an opportunity to talk with others, time permitting. You don't know where a job lead will come from. If you can't talk with some people, get their business cards and leave your resumé. You can always email or write later, and in the meantime, they may see something in your resumé that whets their appetite to call you after they get back to their offices.

15

Being Your
Own Boss

So, you don't want to work for the government as a Department of Defense contractor when you hang up your uniform for the last time. In fact, you don't want to work for anyone or any company, but eating regularly is a habit you and your family have acquired over the years. Now what? Owning your own franchise or doing independent consulting are alternatives you should consider. Both offer a lot of independence but they also can leave you high and dry if you don't do your homework.

Franchises

If you have considerable savings, energy, and patience and are willing to work long hours, you may be a candidate to own your own franchise. As with any unfamiliar business venture, you should do as much research as possible before launching into the franchise held. There are as many failures as there are success stories.

Do extensive market analysis to see how many other franchises are in your area of interest, talk with potential competitors, ask about advertising and support available from the head office, and consider the local available workforce. The Service Corps of Retired Executives, a volunteer organization, is a great source of free wisdom. These retirees have "been there, done that" and can share their management and technical expertise.

A lot of your money will go into the franchise up front, and this may put the financial security of your family at risk. A number of business publications indicated that owners of franchises had to lay out an average of $143,260 to buy a franchise and had an annual pretax net averaging $91,630. Make sure you are willing to commit yourself to a franchise before you commit that money.

Independent Consulting

The great advantage to consulting is you don't have to put up a small fortune to get started, and you will probably make nearly as much as a franchise owner with less responsibility and commitment. It can be hard to get started, but once you do, you can generally work when you want to without the hassles that come with being the boss.

A first step is to assess honestly whether you have a marketable skill that employers are seeking. Trust me, being a "hardworking team leader" is not necessarily a marketable skill. Education and up-to-date, hands on experience in logistics, information management systems, transportation, maintenance operations, or medical systems will be an asset when you seek a consulting position.

As always, look before you leap. Do research through informational interviews with those in your field. Ask them, and yourself, these hard questions: What marketable skills do I really have? How will I attract clients? Am I prepared to take care of taxes, medical benefits, and retirement plans? Will I be happy with a cyclic and uncertain salary? Does my immediate family support my goals?

Finally, think about the goal behind your consulting work. Do you want to learn a new skill to eventually develop a large firm, or to maintain a minimum source of income that will give you the time to pursue other interests?

Whatever course of employment you pursue as a second career, remember, you always work for someone. Either you work for an employer who will handle management and administrative problems or you work for yourself and handle those problems yourself.

16

Getting Fired

T
he U.S. economy has always been on a cyclical ride and at any given time in our history some segments of our economy are on top while other segments are at the bottom. Many of today's workers, and particular those in the military, have never dealt with the challenges associated with losing their jobs and having to start over from scratch. The only reason many of these employees previously changed jobs at all was for a promotion inside or outside their company. But anyone who has kept up with the news knows the economy has slowed during the past year even for the "dot com" and retail segments that, a few years ago, seemed as if they would grow forever.

No one person can really forecast where an economy is going. If they could, Alan Greenspan would have received his own pink slip a long time ago. When significant segments of the economy start to spiral downward, it's only a matter of time before others follow. Obviously, the federal government will take measures to limit the impact of a recession, but if you are an employee with a pink slip in your hand, or suspect one is coming, you need to take your own measures to ease the impact.

It's a rare person who receives a pink slip out of the blue. Signs that one is approaching usually are apparent if you are paying attention. Initially, your industry, profession, or company will send out signals through corporate reports and trade magazines that dividends on stocks are smaller than forecasted, competition has increased, or mergers and consolidations are being discussed. Then, downsizing, restructuring, buyouts or early retirements begin to occur more regularly. Finally, on a more personal level, you may notice that your colleagues seem distant, some of your projects have been given to someone else, or your performance appraisals are not as glowing as they once were, all signs that things are not going well.

Staying prepared is the best way to ensure you are not completely devastated in the event of being downsized. Always keep your resumé updated and continue networking within your professional and social circles. Also, keep your professional and industry certifications current; this is no time to slacken your efforts.

Don't be ashamed to let those in your professional and social circles know you have been laid off. Most of them have probably been through a few setbacks too, and they may have great advice on how to get through your trauma and could know of other job opportunities. Explore the full range of state and federal unemployment job training and financial assistance programs. You paid into these programs your entire working life; now is the time to use them.

If, however, you are suddenly laid off and unsure about what resources are available, remember it is important to face your new challenge with the best attitude. Check into your company's severance package. Most companies offer, at a minimum, two to three week packages with full pay and benefits. Many also offer programs like career counseling and transition training as well as job referrals.

Continue living within your financial means, investing wisely, and saving enough to get you through a few months of unemployment. You will want to be able to afford some time off because finding a job is about finding yourself and discovering your interests and what makes you happy, important questions you must have answered before you can find the next right job for you.

If you are prepared, a pink slip should not break you and may even open an opportunity for you to chase life, not the dollar.

Finding Another Job

This is a wake up call about the current job market. I'm sounding this alarm in light of the many news releases stating how great the opportunities are for finding work. You may have heard the old saying, "Believe half of what you see and none of what you read." Well, it is not as easy to find a job as you might expect in today's job market, so don't take your job search for granted. Put a career transition plan together long before your retirement date and prepare for the challenge.

Every day, the news describes the plight of corporate business' inability to fill jobs with qualified personnel. Every day I counsel bright, articulate officers who have formal education from some of the best schools in the country; have skills in personnel, logistics, and resource management acquired during 20 to 30 years of military service; and are highly decorated and they don't have a job. If unemployment rates are the lowest they've been in 20 years, what's the problem?

The answer lies somewhere between the unskilled, entry level openings the employer has and the mid-level management or supervisory job expectations the officer has. A perfect analogy might be that of the military having difficulty recruiting privates but, generally, no problem retaining senior grade officers and noncommissioned officers. The military is always having problems retaining certain specialized personnel (i.e., aviators, computer programmers), while "soft skill" positions are easy to fill at the required levels.

Comparing civilian and military opportunities offers a great lesson in supply and demand. If you have information technology skills, employers are looking for you. If you have a degree in sociology, like me, your job-hunting work is cut out for you, even in a robust economy. It may take longer than you expected to find a great job, and when you do, it most likely will not be at the same pay grade, responsibility, and status you enjoyed on active duty.

There are a number of options for the officer and noncommissioned officer with a soft skill degree (psychology, political science, history, etcetera) who has performed tough, challenging jobs in the field. Your first option is to get certified in a skill that has high demand: anything in information technology, computers, Microsoft network systems, and so on. But do not think that dashing off to the local university to get another degree is the answer to finding a job. Employers want people with hands on experience that can make systems work immediately without long and expensive training. Do your homework, and make sure not only that there is a demand for the certification you are pursuing but also that you have the applicable hands on skills to go with the education.

Other certification possibilities include becoming a paralegal, real estate manager, financial planner, meeting planner, or certified public accountant. You might consider accepting an entry-level management trainee position just to get your foot in the door of an organization. This does not necessarily mean flipping hamburgers; you can start at a local Home Depot or Sears, take a position in retail sales, teach at the secondary level, or take a staff position at a small college or high school.

Finally, networking is the activity most likely to get you a job. It will give you the opportunity to sell yourself before you have to sell your military skills to an employer who has never served in the military. In a non-technical environment, enthusiasm, willingness to work hard, discipline, and chemistry far outweigh education. You ultimately will succeed in your job search but do not take anything for granted.

17

Thank-You Notes

hank you. These are probably two of the most important words in any language but they are not used often enough. Today's fast paced society, and particularly the business community, encourages people to hustle, to get the job done, and to make a profit often at the expense of civility among the people involved. Don't let this happen to you during your transition to the corporate world. Take with you those social skills you have acquired in the military community and continue to practice them in the corporate world; those skills will pay great dividends.

I believe the formality of social functions within the military community encourages greater use of etiquette and attention to social graces than is frequently observed in many business organizations. These same social graces, such as allowing women to precede men, opening doors for others, standing up when you are introduced, walking closest to the traffic on the sidewalk (for a man, when a woman is present), and routinely showing appreciation by thanking others, will serve you well during your transition and subsequent employment. Don't be lulled into complacency just because others are not attentive to these social requirements.

Many of you will say, "So what's new? I routinely thank others and I am aware of most social etiquette." This may be true, however, how many of you have written a thank-you note lately? This is not a big deal and it's more reflective of your self-discipline than of your ability to write.

A thank-you note during your transition can convey far more to a potential employer than you might believe. A note gives you the opportunity to highlight information that wasn't discussed during the interview, express thanks for networking leads, or show appreciation for other information you received that will be useful during your job search. You can also use the thank-you note as a means to keep the doors open to other potential job opportunities and networking sources, to ask for additional information, or even as an appropriate way to gracefully say no to a job offer.

What are some of the key elements of a thank-you note? Hopefully, you caught the word "note," because that is what it should be—a note—not a book. Start with the full, formal, and correctly spelled name of the person you are thanking. (Do not strike through the last name and add a handwritten "Joe" or "Judy," as is frequently done in the military.) Typically, a typed one-page letter on bond paper is the most professional approach. This is appropriate for the hiring official or senior person in the organization. It may include every member of the interview panel.

On a few occasions use your intuition. A handwritten note card is just as appropriate if it is for an administrative person who provided information or other assistance during your visit. Sometimes a phone call also may be the most effective way to say thank you, especially if the recipient is a friend. This is your decision. A good rule of thumb is the more senior the person, the more formal the manner and method of the thank-you.

Be brief and to the point, with a positive tone, even if you are writing the note after finding out you did not get the job. In this instance the employer may think, "well she may fit in another department" and forward your resumé to that department for consideration. You just never know.

Timeliness is also important. Ideally, you should send a thank-you note within 24 hours, so your visit or interview will be fresh in the minds of those who helped you or conducted the interview. A thank you note is not a tough project. There are thousands of examples in the reference books if you need them. Ideally, you should have the envelope addressed and the letter drafted before the interview or meeting so that with minor changes you can finalize the note and get it in the mail immediately. Faxing is another increasingly popular way to get your thank-you note to its recipient quickly.

Finally, and with care not to overdo it, you may want to follow your note with a personal phone call a few days after you mail the thank-you note. You might say something like, "Ms. Jones, you probably received my thank-you note but I want to thank you again for taking time to meet with me last week. I also want to see if I might provide any other information." Afterward you can only stand by and see what happens. It can take anywhere from a few days to a few months before a final decision is made.

18

Summary

I n my many years of lecturing around the world for the Military Officers Association of America, I have gathered several pearls of wisdom about the career transition process. There are few original thoughts left in life, so I take no credit for any "original" infomation in the preceeding pages (and I've credited the source if I know it).

Focus on the type of work you want to do, not the title. You will never again be a commander, colonel, or captain; it's Bob or Sue, Mr. or Ms. No one cares what rank you had while on active duty. They want to know what you can do for their organization now. You will put all of your time, money, blood, sweat, and tears into career transition planning and execution and can only hope you are successful. You are pretty much on your own once you hang up your service uniform. What happens to you will be solely the result of your efforts.

"You have a better chance of winning the lottery than you do of getting a job out of the newspaper. The more senior the position, the less likely it will be filled through want ads or the Internet." (From *What Color Is Your Parachute?* written by Richard Boles.) On the other hand, most of us have played the lottery once or twice, so go ahead and use the newspaper and the Internet as part of your career transition program—you never know what may happen. No one ever gotten a job sitting at home watching television. Get out of the house and meet people who are working through aggressive networking. (Work defined: "toil, labor, drudgery, travail"—*The American Heritage Dictionary*.)

In uniform, you have a profession and an avocation. You share a whole range of bonds that no other profession has. If you like (or even half like) being in uniform, stay in until they throw you out.

"Chemistry" is more important than knowledge in getting a job. Only 30 percent (if that) of an interview is based on substance. Your personality and who you know are far more important than what you know.

Don't start your career transition in debt. If you have more than one credit card or more than $1,000 in credit card debt, you have potential problems.

"I can teach leaders about high technology a lot (more easily) than I can teach technology people about leadership"—(Bernard Bailey, U.S. Naval Academy graduate and IBM executive.) You have a lot of marketable "soft" skills that many employers would love to have in their organization. Learn to explain them to civilians.

Success is what you define it to be. Don't let others define success for you unless they are paying your mortgage and other bills. As you start the career transition process, consider what makes you and your family happy. Don't worry about what the Jones will think.

During your career in uniform, you've been sent all over the world, sometimes with very little training or notice and often with minimal support. You landed on your feet, figured out what had to be done, and made things happen. The career transition process may be something you have never done before, but with a good attitude, you will again land on your feet. You are not alone, nor are you the first to leave the security of being in uniform. You will find success in whatever you choose to do.

A P P E N D I X   A

Resumé Examples

There are few original thoughts left in life so do not spend your life trying to write the one and only original resumé. Attached below are resumés that others have used during their career transition. Find a few examples that best describes your career and "fine tune" them to fit your new career objectives. These are, obviously, not perfect resumés. The only perfect resumé is the one that gets you an interview.

Combination Resumé

John K. Doe, Sr.
1000 Apple Lane Ct, Halifax VA 22513
(Home) (803) 951-7290 (Work) (803) 340-2211 (FAX) (803) 420-6966
(Home) doej275@aol.com (Work) john.doe@army.pentagon.mil

Objective
Seeking a Program Managerial position in law enforcement, force protection or security operations.

Professional Qualifications
Over 28 years of US Army leadership and management experience in the fields of security operations, law enforcement, anti-terrorism, force protection, policy development and personnel training. Decisive and confident team builder who effectively interacts with, motivates, and mentors a diverse work force, creating cooperative partnerships among all levels inside and outside of the organization. Top Secret Security Clearance.
  • Commanded military police organizations at virtually every level in the Army,

including leading security and law enforcement forces in multinational operations abroad in both peaceful and hostile environments.

- Extensive national level analytical, policy development and operational experience across a wide spectrum of challenges ranging from security of major Army installations to strategic planning and operations for the Department of Defense. Chief operations officer for Army programs exceeding $1.7 billion.

Career Highlights

2002-Present. **Chief, Operations Officer**
Security, Force Protection and Law Enforcement Division, Headquarters, Department of the Army, Pentagon.

- Advised Army senior leaders on law enforcement, anti-terrorism matters and strategies regarding homeland defense, to include deployment of military police units throughout the world. Supervised over 30 special event security requirements for flag officer level dignitaries in the National Capital Region.
- Synchronized the efforts of a division of 70 personnel, ensuring Congress is kept informed on the Army's programmatic and policy requirements for law enforcement, anti-terrorism, physical security, military working dogs, corrections, and enemy prisoner of war/detainee operations. Planned and developed military police rotation, employment, and force reduction initiatives, resulting in major manpower savings for the Army.
- Directly involved in developing law enforcement strategic vision, programs, and training requirements for the Military Police Corps. Developed structure and policy redesign in establishing the Office of the Provost Marshal General for the Army. Coordinated training curriculum for DOD police, guards, and military police leaders. Lectured as a recognized subject matter expert on Provost Marshal Operations and Public Safety.

2000-2002. **Commander and Director**
Military Police Battalion and Public Safety Business Center, Fort Campbell, Kentucky.

- Commanded one of the Army's largest Military Police Battalions with over 1200 assigned personnel performing a wide-variety of diverse security and law enforcement activities. Trained and deployed military police to support the global war on terrorism in Afghanistan and Guantanamo Bay, Cuba.
- Directed law enforcement, security, and fire emergency response operations in support of a community of 60,000 personnel. Managed an annual budget of $57 million. Implemented community relations programs such as community policing; well-being and safety programs by developing an enhanced emergency notification system for fire, police and medical responders, saving thousands of dollars annually as well as reducing response time. Special Reaction Team was noted as one of the best in the Military Police Corps during an installation terrorism counteraction assessment program conducted by an independent evaluation.

- Senior leader of the Military Police Task Force conducting law enforcement, peacekeeping and security in a multinational and hostile environment in Kosovo. Planned and coordinated security for visiting dignitaries to include President and Mrs. Bush. Chaired a successful joint security-working group with 14 sector military and non-governmental organizations. Redesigned the detention facility, enhancing safety of detainees and guards. Reduced the escape rate to zero.

1998-2000. Provost Marshal
Public Safety Business Center, Fort Bragg, North Carolina.
- Directed and managed law enforcement and force protection for an installation over 160,000 acres with a population of 150,000 personnel. Managed an annual budget of $1.5 million. Focused on community response, reducing formal complaints against police by 43%. Conducted traffic and patrol distribution studies, the first documented in 10 years, facilitating changes of traffic patterns and targeting of high crime areas.
- Supervised the police station and staff of 170 personnel daily in the conduct of law enforcement, crime prevention, confinement, kennels, investigations and police administration. Instituted proactive programs such as selective enforcement operations to target specific criminal activities. Reduced thefts of automobiles, motorcycles and vending machines. Targeted gang and illegal drug activities, resulting in increased apprehensions of criminals and reduction in crime. Special Reaction Team won "Top Gun" at the Military Police School and received honors in a variety of competitions, as well as successfully conducting a series of raids against armed and unarmed assailants with no injuries.

1997-1998. Executive and Operations Officer
Military Police Brigade, Fort Bragg, North Carolina.
- Managed and coordinated staff support and leadership functions required to organize, equip, man, maintain, train and deploy the force. Coordinated contingency planning for five subordinate battalions of 2500 personnel. Refined and developed training packages, which improved readiness, focusing on mission essential training objectives by specific task, operating conditions and establishment of training standards. Assisted in the design and establishment of both an indoor and outdoor range with a variety of non-standard firing positions to enhance individual and team proficiency using challenging scenarios and obstacles.
- Directed and supervised administrative, medical, logistics, security, intelligence, operations, maintenance, and training support. Mentored and developed a junior staff who successfully assisted in training and deploying units to conduct peacekeeping operations in Bosnia.

1995-1997. Executive and Operations Officer/Deputy Division Provost
Marshal Military Police Battalion, Fort Drum, New York.
- Managed and coordinated staff support and leadership functions required to

organize, equip, man, maintain, train and deploy the force for an installation over 107,265 acres in size, serving a population of 25,000 personnel. Ensured the proper care, maintenance, and accountability of facilities and equipment in excess of $26 million for three subordinate units of 370 personnel. Recognized as having one of the top two executed budgets on the installation.

• Directed and supervised administrative, logistics, security, intelligence, operations, maintenance, and training support. Selected as chief observer/controller for evaluating eight subordinate units, which were successfully deployed for peacekeeping duties in the Sinai and Panama.

Civilian Education

MPA, Public Administration, Jacksonville State University
BS, Criminal Justice, Fayetteville State University

Military Education

Senior Officer Legal Orientation Course, Charlottesville, VA
Defense Strategy Course, US Army War College, Carlisle Barracks, PA
Air Force Command and General Staff College, Maxwell AFB
Canadian Land Forces Command and Staff College, Kingston, Canada
Combined Arms and Services Staff College, Fort Leavenworth, KS
Terrorism Counteraction Course, Fort McClellan, AL
Installation Provost Marshal Course, Fort McClellan, AL
Special Reaction Team Course, Fort McClellan, AL
Military Police Officer Advanced Course, Fort McClellan, AL

Functional Resumé

Name
Address
Telephones
Emails

Objective: Supervisor/Director of Supply Chain, Material Logistics, Inventory Control for a manufacturing company

Summary: Over 20 years of extensive, hands-on experience in all areas of Supply, Financial, and Material Logistics management. Documented superior results while achieving the highest levels of productivity and efficiency. An experienced teacher and manager with exceptional administrative and customer service skills. Relevant skills include:

- Inventory Management
- Facilities Management
- Financial Management
- Customer Service
- Program/Configuration
- Instructor

Accomplishments:

- **Inventory:** Reduced inventory by 55% and financial investment by 33% while increasing Management material availability from 82% to 91% by eliminating unnecessary stock levels, coordinating more closely with suppliers to be more responsive and streamlining stock replenishment procedures.
- **Facilities:** Planned and executed a complete re-warehouse plan that reduced the real Management estate footprint from 1.5 million square feet to 300,000 square feet with a corresponding decrease in fixed maintenance and utility costs of 50%.
- **Financial:** Developed a ship wide standard financial management plan template for Management shipboard use. Combining zero-based budget analysis and historical data from previous classes of ships, it established a centralized funding strategy for cross department and unique programs such as Information Technology and Hazardous Material management.
- **Customer Service:** Improved the quality and timeliness of Military Sealift Command (MSC) Service Combat Logistics Force shipboard loadouts. This improvement led to better service for all supported activities throughout the Western Pacific, Indian Ocean, and Persian Gulf areas.
- **Program Management:** Identified and corrected over 8000 material, inventory and configuration discrepancies in logistics support plans. Developed an efficient and logical plan of corrective actions to ensure all database entries were correct which allowed the logistics support deliverables to be on time.
- **Instructor:** Instructed over 400 newly commissioned Naval Supply Corps officers in basic shipboard supply, financial, logistics, disbursing, sales, and inventory management. Developed a unique training plan for the US Coast Guard that was accredited by the Southern Association of Colleges and Schools and judged to be one of the best of 400 schools inspected.

Work History:

Program Management Analyst
Defense Threat Reduction Agency, Fort Belvoir VA   *3 years*
  • Responsible for program formulation, defense, review and analysis of a widely diverse portfolio of warfighter support programs approximating $2 billion per year in operating and investment resources.

Material, Storage, and Distribution Manager
Fleet and Industrial Supply Center, Agana, Guam   *2 years*
  • Planned and managed the efficient storage, inventory, packaging, and distribution of material. Managed 122 personnel in the processing of over 100,000 receipts and issues annually. Developed and controlled an annual operating budget of $2 million.

Requirements Planning Manager
Fleet and Industrial Supply Center, Agana, Guam   *2 years*
  • Managed the material/inventory requirements planning process for 100,000 items of inventory and a retail store with $1.5 million in sales. Directed computer analysis in the development and analysis of automated tools and programs. Supervised 25 personnel in the performance of their duties.

Staff Instructor
Navy Supply Corps School, Athens, GA   *2 years*
  • Instructed over 200 newly commissioned Naval officers in all areas of shipboard supply and logistics support. Developed a specific Supply Management course for the U. S. Coast Guard and served as course coordinator during successful accreditation process.
  • Other work assignments: Logistics manager for a 400-person organization with 20,000 line items valued at over $5 million and a $2 million dollar operating budget. Distribution and retail manager for a 500-person organization with 25,000 line items of inventory valued at $21 million and $150K in annual sales and an operating budget of $4 million.

Education:
M.S. Management (Inventory Control/Material Logistics),
    Naval Post Graduate School, Monterey, CA
M.S. Human Resource Management, National University, San Diego, CA.
B.A., Business Administration and Economics, Carroll College

Combination Resumé

Name
Address
Telephones
Emails

Objective:
Seeking to manage a large-scale Industrial facility plant

Summary of Qualifications:
Over 10 years of high-level physical plant administration experience. An outstanding ability in planning, organizing, directing and controlling resources in large-scale facility management programs. Strong interpersonal skills and analytical abilities demonstrated in positions of increasing responsibility over facility planning and management, real property services, and maintenance/repair activities in many varied domestic and international environments.

Education:
M.S., Systems Management, University of Southern California
B.S., Industrial and Systems Engineering, The Ohio State University

Selected Accomplishments:
- **Facility Planning:** Managed short and long-range facility and infrastructure capital investment programs valued over $400M annually. Directed staff of 22 in developing capital investment strategies for real estate assets with a plant value exceeding $15 billion at 12 United States and 21 overseas locations. Recognized for implementation and management of aggressive $1.2 billion overseas facilities/infrastructure maintenance/repair and construction program which resulted in a greatly improved Department of Defense worldwide air transportation system.
- **Leadership:** Directed professional engineering and facilities management services supporting 14,000 people at 104 locations worldwide. Developed and executed multiple source funding strategy for the facility project program valued at $20 million annually. Oversaw environmental policy and guidance. Recognized for performance of outstanding service as the chief executive officer providing facility services to the Air Force, Army, Navy, National Security Service, and other US intelligence agencies.
- **Real Property Services:** Directed 128 personnel in the acquisition, construction, operation, and maintenance/repair of utility and road systems and 365 buildings on a 516-acre estate with an annual budget of $12 million. Responsible for full spectrum of real property services including water, waste, trash collection, grounds' maintenance, pest control, custodial services, and the maintenance, lighting and snow removal of roads and walkways. Managed a

413 unit family housing complex. Developed the Air Force's first full-time environmental protection office in Europe cited for the best Air Force recycling program in England—reduced annual trash collection costs by $20,000. Established unit's first Total Quality Management Program and led continuous improvement efforts among other base organizations. Recognized for the best Air Force housing management, fire protection, and environmental operations in Europe.

- **Process Improvement:** Developed, planned, and conducted Air Force-wide functional management studies, and other investigative/research activities—each was completed on schedule and under budget. Planned worldwide study of Air Force's environmental program resulting in a detailed management assessment at $60,000 under budget. Planned, organized and conducted thorough study of world-wide Air Force Real Property Maintenance budgeting process—cited for innovative recommendations that provided Air Force senior leaders the management tools to more effectively direct and control the entire $2.5 billion annual operations and maintenance budget.

- **Project Management:** Managed the execution of construction projects ranging in cost from $25,000 to $3 million, coordinating the efforts of 75 engineers and numerous Architect/Engineer firms. Established the procedures, practices, and policies that cut project completion time in half while manpower was being reduced by 1/3. Nearly $40 million in projects over a two-year period completed on schedule and at cost.

Experience:

Positions of increasing responsibility as an Industrial Engineer in the administration of the US Air Force's physical plant:

2002-Present   **Deputy Chief, Planning and Programming Division.**
Directorate of Civil Engineering, Headquarters Air Mobility Command Scott AFB, Illinois

2000-1997   **Director, Civil Engineering**
Headquarters Air Intelligence Agency, Lackland AFB, Texas

1997-1995   **Commander and Base Civil Engineer**
774 Civil Engineering Squadron, Royal Air Force, Chicksands, England

1995-1993   **Civil Engineering Inspection Team Chief**
Headquarters Air Force Inspection and Safety Center, Norton AFB, California

Affiliations:

Institute of Industrial Engineers
Society of American Military Engineers

Functional Resumé

Name
Address
Telephones
Emails

Objective: Supervisor/Director of Supply Chain, Material Logistics, Inventory Control for a Manufacturing company

Summary: Over 20 years of extensive, hands-on experience in all areas of Supply, Financial, and Material Logistics management. Documented superior results while achieving the highest levels of productivity and efficiency. An experienced teacher and manager with exceptional administrative and customer service skills. Relevant skills include:

- Inventory Management
- Facilities Management
- Financial Management
- Customer Service
- Program/Configuration
- Instructor

Accomplishments

- **Inventory:** Reduced inventory by 55% and financial investment by 33% while increasing Management material availability from 82% to 91% by eliminating unnecessary stock levels, coordinating more closely with suppliers to be more responsive and streamlining stock replenishment procedures.
- **Facilities:** Planned and executed a complete re-warehouse plan that reduced the real Management estate footprint from 1.5 million square feet to 300,000 square feet with a corresponding decrease in fixed maintenance and utility costs of 50%.
- **Financial:** Developed a shipwide standard financial management plan template for Management shipboard use. Combining zero-based budget analysis and historical data from previous classes of ships, it established a centralized funding strategy for cross department and unique programs such as Information Technology and Hazardous Material management.
- **Customer:** Improved the quality and timeliness of Military Sealift Command (MSC) Service Combat Logistics Force shipboard loadouts. This improvement led to better service for all supported activities throughout the Western Pacific, Indian Ocean, and Persian Gulf areas.
- **Program:** Identified and corrected over 8000 material, inventory and configuration discrepancies in logistics support plans. Developed an efficient and logical Management plan of corrective actions to ensure all database entries were correct which allowed the logistics support deliverables to be on time.
- **Instructor:** Instructed over 400 newly commissioned Naval Supply Corps officers in basic shipboard supply, financial, logistics, disbursing, sales, and inventory management. Developed a unique training plan for the US Coast Guard that was accredited by the Southern Association of Colleges and Schools and judged to be one of the best of 400 schools inspected.

Work History

Program Management Analyst
Defense Threat Reduction Agency, Fort Belvoir, VA   *3 years*
  • Responsible for program formulation, defense, review and analysis of a widely diverse portfolio of warfighter support programs approximating $2 billion per year in operating and investment resources.

Material, Storage, and Distribution Manager
Fleet and Industrial Supply Center, Agana, Guam   *2 years*
  • Planned and managed the efficient storage, inventory, packaging, and distribution of material. Managed 122 personnel in the processing of over 100,000 receipts and issues annually. Developed and controlled an annual operating budget of $2 million.

Requirements Planning Manager
Fleet and Industrial Supply Center, Agana, Guam   *2 years*
  • Managed the material/inventory requirements planning process for 100,000 items of inventory and a retail store with $1.5 million in sales. Directed computer analysis in the development and analysis of automated tools and programs. Supervised 25 personnel in the performance of their duties.

Staff Instructor
Navy Supply Corps School, Athens, GA   *2 years*
  • Instructed over 200 newly commissioned Naval officers in all areas of shipboard supply and logistics support. Developed a specific Supply Management course for the U. S. Coast Guard and served as course coordinator during successful accreditation process.
  • Other work assignments include: Logistics manager for a 400-person organization with 20,000 line items valued at over $5 million and a $2 million dollar operating budget. Distribution and retail manager for a 500-person organization with 25,000 line items of inventory valued at $21 million and $150K in annual sales and an operating budget of $4 million.

Education

M. S. Management (Inventory Control/Material Logistics), Naval Post Graduate School, Monterey, CA
M.S. Human Resource Management, National University, San Diego, CA
B.A., Business Administration/Economics, Carroll College, Waukesha, WI

Functional Resumé

Name
Address
Telephones
Emails

Objective: Financial Institution Manager within the financial services industry.

Summary of Experience: Over sixteen years experience and success in senior executive leadership positions, managing organizations of up to 650 diverse men and women. Strengths include: financial management, customer relations, resource management, administration, public presentations, and planning.

Education:
B.A., Economics, Old Dominion University
M.P.A., Public Administration, University of West Florida American
   Institute of Banking

Accomplishments:
- **Financial:** Planned and administered annual budgets up to $5.2 million, allocating funds among competing programs and departments. Results: saved $500 thousand without eliminating or degrading existing programs; rehabilitated office spaces with savings, boosting both morale and productivity.
- **Customer:** Extensive training and practical experience in Total Quality Management and customer satisfaction. Proven tactful demeanor has contributed substantially to problem solving, enhancement of workplace harmony, productivity, employee efficiency and improved communications.
- **Resource Management:** As head of training organization, corrected nine week student backlog. Results: saved $17.5 thousand dollars, monthly, in travel funds. Zero backlogs contributed to five percent higher manning in ultimate work units.
- **Administration:** Coordinated work efforts for staff of 500 diverse men and women. Improved administrative process by introduction of computer routing and tracking of paperwork, reducing required reports and cutting task attainment time by fifty percent.
- **Public Presentations:** Planned, created, and delivered polished presentations on a myriad of subjects, to include policy, leadership, finance, and technical matters. Comfortable in diverse forums.
- **Planning:** Originated and refined comprehensive relocation plan for two units (400 employees) and shutdown plan for three others units. Results: Saved $6 million in operations expense by accelerating actions within fiscal year.

Work History:

Training & Readiness Officer, Naval Air Force, Atlantic   Norfolk, VA
- Responsible for all aspects of training and readiness for three functional helicopter wings; 22 squadrons with 216 aircraft and 5500 personnel.

Executive Assistant, U.S. Atlantic Command   Norfolk, VA
- Directed the work efforts of staff of 500 diverse men and women. Managed building renovation and staff relocation.
- Developed policies and procedures for coordination of staff work efforts towards accomplishment of organizational goals.

Student, National War College   Washington, D.C.
- Graduate level curriculum of national security policy formulation for select senior military officers and senior executive officials from federal agencies.

Squadron Commander   Mayport, FL
- Head of 650-person organization, responsible for training U.S. and foreign military pilots and aircraft maintenance technicians. Created student training and aircraft maintenance plans, formulated and managed $5.2 million budget, counseled students and instructors, developed curriculum, and conducted numerous public presentations.

Affiliations:

National Honor Society for Public Administration American Association of Individual Investors

Functional Resumé

Name
Address
Telephones
Emails

Objective: Seeking a position as Director of Planning and Administration

Career History: Eighteen years of documented success in program, resource and financial management, and in administration, telecommunications, and computer based information handling systems. Hands-on player in developing industry-wide policies and programs involving manufacturing and performance standards, industry promotion, legislative and regulatory issues, and environmental affairs.

Education: B.S. Finance, University of Kentucky

Planning and Administration:
- Directed 16 industry groups developing manufacturing and performance Standards. Coordinate multi-departmental technical functions at all levels of management in a matrix management environment.
- Initiated and actively managed short term (one-year) and long term (five-year) schedules used in standards development. Served on two international committees harmonizing U.S. and European performance standards in support of global markets. Ensured compliance with requirements of other standards writing, safety monitoring and testing organizations. Worked closely with federal and state agencies on industry regulatory issues.

Financial Management:
- Developed/executed financial and personnel management plans for automation management and electronics groups to track $3.5 million budget. Created program budget and manhour tracking/reporting system. Allowed senior management to instantly analyze effects of budgetary and personnel actions and develop "what-if" scenarios. System was implemented organization-wide.
- Controlled $400K annual budget. Reduced expenses by 7.8% over a three-year period by establishing program priorities and instituting tighter controls on travel and conferences.

Operations Management:
- Directly supervised telecommunications and computer support services to the U.S. Navy as senior administrator for financial management and budgeting. Developed organizational restructuring plan to support a down-sizing. Result was an overall 3.5% reduction in manpower and 12% reduction in central support staff over a two-year period.

- Performed analytic studies of funding and staffing requirements to ensure optimization of resources provided to field activities. Result was flexible, prioritized support plan that balanced resources against need as communications configurations, hardware, and tasking evolved. Reduced staffing shortfalls by 5%. Supervised training plans and programs. Prepared budget estimates. Designed and instituted automated tracking system to monitor program development and expenditure of funds. Realized 6% savings in training budget.

Resource Management:
- Directed maintenance of electronic equipment and facilities valued at $16 million and supported by an annual budget of $375K. Over a three-year period, achieved 9.5% savings on repair parts by acquiring and installing on-site micro-miniature repair station. Team leader in designing replacement facility, including architectural review, space planning, equipment installation, and furniture acquisition.
- Functional Manager for major computer upgrade. Specified requirements and budgets for acquisition and installation of state-of-the-art equipment and software. Developed immediate and long-range (two and one-half years) phased equipment replacement program.

Business Development:
- Marketed national trade association to prospective members. Developed industry promotion plans. Organized and staffed conferences to promote industry and conducted extensive surveys and interviews with clients and association members. Payoff was a 6% increase in industry membership and 58% increase in client participation over a two-year period, despite recessionary market conditions.
- Developed and administered industry-wide policies and programs dealing with issues such as technology transfer, legislative and environmental affairs, advancement of technical education, competitiveness of U.S. industry, global markets, and trade barriers.

Communication Skills:
- Extensive background in preparing written documents, reports, and visual displays for presentation to and use by senior executives in marketing and budgeting issues. Coordinated presentations of automation and electronics program managers. Computer literate. Experienced with major software programs (WordPerfect, Lotus, and others) written for MS/DOS IBM PC and compatibles.

Chronological Resumé

Jerry C. Willis
6223 East Gate Lane, Norfolk, VA 23452
(757) 614-3993,jerry,willis@saalt.army.mil
Home (757) 922-8147, jwillis273@aol.com

Objective
Seeking senior program management position involving systems acquisition or logistics.

Summary of Qualifications
Over 29 years experience as an Army officer, 16 within the Army Acquisition field. Certified Acquisition Professional with documented success and practical experience in leadership, team building, planning and program management with diversified organizations. Expertise includes Systems Acquisition, Project Management, Logistics Management and Program Integration.

Education & Special Qualifications
MBA, Logistics Management, Florida Tech
BS, Mechanical Engineering, University of Toledo
Program Manager's & Executive Program Manager's Course, Defense Systems
    Management College
Senior Service College, Army War College
Top Secret, SBI/SCI, security clearance
DAWIA Certified Level III: Program Management

Experience
Director of Program Integration, Assistant Secretary of the Army,
Acquisition, Logistics and Technology
Pentagon, Washington, D.C.   *2001-Present*
- Manage program integration activities at the Headquarters, Department of the Army, level across all Army acquisition programs funded annually at over $20 billion. Provide acquisition oversight of the Army's major integration initiatives involving systems of systems development, system architectures and synchronization of multiple interoperating acquisition programs. Coordinate the operations of the various directorates within the Secretariat and provide interface across the Army staff. Ensure programs are effectively funded each year within the DoD and Congressional budget process. Responsible for long range funding plans. Use extensive program management experience to mentor current program managers in developing executable programs. Advise Army Acquisition Executive and other Army leadership on key acquisition issues. Manage staff of twenty-six professionals.

Project Manager of the Department of Defense
Mobile Electric Power Program, Fort Belvoir, VA   *1997-2001*
- Directed a $1 billion plus joint service program developing and acquiring state-of-the-art mobile electric power generation systems. Responsible for five diverse major programs spanning early development to production and fielding. Coordinated all phases of the programs with industrial partners and joint service customers. Awarded seven major new contracts. Infused new technology into the military's power generation fleet: digital controls and displays, light weight engines, improved electronics. Managed program, engineering, logistics, contracting and support staff of 250 professionals.
- Exceeded customer requirements for portable power generators by reducing weight 53%, improving reliability 43% and lowering overall operating costs.
- Developed acquisition strategy for a new cutting-edge technology approach to the medium power range of generators, convinced leadership to fund development and produced prototype design to prove out approach. Resulted in 50% reduction in number of configurations and in weight, while providing 25% reduction in fuel consumption and total operating costs.

Army War College, Carlisle, PA   *1996-1997*

Product Manager for Forward Area Air Defense
Sensors Program, Redstone Arsenal, AL   *1993-1996*
- Directed a $500 million plus Army program developing and acquiring advanced Air Defense sensors. Responsible for cost, schedule, and performance of developmental and production systems. Successfully transitioned the Sentinel Radar, a major acquisition program under Office of the Secretary of Defense (OSD) oversight, through testing and into production. Established a foreign military sales agreement with Turkey. Managed staff of 80 professionals. Modified production strategy reducing costs by 25%. Reduced core personnel strength by 20%, while maintaining performance.

Assistant Project Manager for Brilliant Anti-armor Tank
Submunition Program, Redstone Arsenal, AL   *1991-1993*
- Led government production planning for the Army's most advanced submunition. Managed government and contractor activities including prime contractor with six major subcontractors in the successful on schedule delivery of the initial prototype hardware for first development flight test. Established production planning for a $2 billion program.

Department of the Army Systems Coordinator, Assistant Secretary of the Army, Research, Development and Acquisition
Pentagon, Washington, D.C.   *1988-1991*
- Coordinated Washington D.C. operations for three tactical missile major defense acquisition programs totaling $10 billion. Provided Army Secretariat

level analysis of program planning, programming and budget. Defended programs at Army, OSD, and Congressional level. Successfully led operations of each the three programs through a major OSD acquisition milestone.

Commander
108th Brigade Support Element   *1986-1987*
- Senior executive for a dispersed logistics organization operating across central Germany in support of U.S. Army Air Defense operations. Provided maintenance and supply support around the clock to supported units, consistently maintaining a 90% operational availability rate. Led an organization of over 400 personnel. Attended Command And General Staff College prior to follow-on position.

Same Resumé As Preceding but In a Combination Format

(Note highlighting of key results)

Jerry C. Willis

6223 East Gate Lane, Norfolk, VA 23452
(757) 614-3993, jerry.willis@saalt.army.mil
Home (757) 922-8147, jwillis273@aol.com

Objective

Seeking senior program management position involving systems acquisition or logistics.

Summary of Qualifications

Over 29 years experience as an Army officer, 16 within the Army Acquisition field. Certified Acquisition Professional with documented success and practical experience in leadership, team building, planning and program management with diversified organizations. Expertise includes Systems Acquisition, Project Management, Resource Management, Logistics Operations and Program Integration.

Education & Special Qualifications

MBA, Logistics Management, Florida Tech
BS, Mechanical Engineering, University of Toledo
Program Manager's & Executive Program Manager's Course, Defense Systems
    Management College
Senior Service College, Army War College
Top Secret, SBI/SCI, security clearance
DAWIA Certified Level III: Program Management

Professional Achievements

Program Management

- Certified Program Management
- 16 years experience with 4 years as a project manager and 3 years as a product manager.
- Directed a $1 billion plus joint service program developing and acquiring state-of-the-art mobile electric power generation systems. Responsible for five diverse major programs spanning early development to production and fielding.
- Exceeded customer requirements for portable power generators by reducing weight 53%, improving reliability 43% and lowering overall operating costs.
- Directed a $500 million plus Army program developing and acquiring advanced Air Defense sensors. Successfully transitioned the Sentinel Radar, a major acquisition program under Office of the Secretary of Defense (OSD) oversight, through testing and into production.
- Modified production strategy reducing costs by 25% and core personnel

strength by 20%, while maintaining performance.
- Awarded eight major new contracts.
- Led early government production planning for the Army's most advanced submunition.

Resource Management
- 16 years experience with the Department of Defense Planning, Programming and Budgeting System (PPBS).
- Developed program and budget for billion-dollar power generation program. Defended program at Army and OSD.
- Provided Army Secretariat level analysis of program planning, programming and budget for three tactical missile programs totaling $10 billion. Defended programs at Army, OSD, and Congressional level.

Logistic Operations
- Over 9 years experience with several senior executive positions.
- Directed operations of a dispersed logistics organization operating across central Germany in support of U.S. Army Air Defense operations. Provided maintenance and supply support around the clock to supported units, consistently maintaining a 90% operational availability rate.
- Increased availability of backup maintenance systems by 50%, while increasing critical part availability by 100%.

Program Integration
- Managed program integration activities at the Headquarters, Department of the Army, level across all Army acquisition programs funded annually at over $20 billion.
- Provided acquisition oversight of the Army's major integration initiatives involving systems of systems development, system architectures and synchronization of multiple interoperating acquisition programs.
- Integrated operations of the various directorates within the Army Secretariat and provided interface across the Army staff.
- Led operations in support of integrated fielding of multiple acquisition programs.

Experience

Director of Program Integration, Assistant SECARMY Acquisition,   2001-2003
Logistics and Technology, Pentagon

Project Manager of the Department of Defense   1997-2001
Mobile Electric Power Program, Fort Belvoir, VA

Product Manager for Forward Area Air Defense Sensors Program   1993-1996
Redstone Arsenal, AL

Assistant Project Manager for Brilliant   1991-1993
Anti-armor Submunition Program, Redstone Arsenal, AL

Department of the Army Systems Coordinator   1988-1991
Assistant Secretary of the Army, Research, Development and
Acquisition, Pentagon, Washington, D.C.

Commander, 108th Brigade Support Element,   1986-1987
Worms, Germany

Various Logistics and Air Defense Assignments   1973-1986

A p p e n d i x   B

Cover Letter Example

Cover letters should be an individual and personal effort; however, in most cases are rarely read by the hiring authority. If you are busy, as most people are, why read the same information twice. The relevant information should be in your resumé. If you feel compelled to write a cover letter, keep it to one page or half a page, if possible. Address it to the hiring authority, be positive, use action verbs, address the specific job requirement and ask for an interview. The following Cover letter make help you get started:

Gerald L. Crews
7419 Running Fast Court
Springfield, VA 22145
Email:

703 922-8764 (H)
703 838-1234 (O)
757 420-4315 (C)

July 7, 2004

Ms Barbara Smith
XYZ Enterprises
1850 Hargrove Lane
Alexandria VA 2317

Dear Ms Smith:

I am very much interested in seeking a challenging career opportunity in a project or administrative management position with your organization. Mr. Bob Johnston, in your logistics operations division, suggested that I contact you regarding job announcement number 2152 seeking a supply management specialist.

With a B.A. in Finance and M.S. in International Relations, I have over 20 years of Military related experience in a wide range of supply, maintenance, and facility management programs. I have command, operational, and high-level staff experience in many aspects of inter-service and inter-governmental agency programs involved in management of logistics operations. Please see the attached resumé for additional details.

I bring to any organization a wealth of leadership experience in building relationships, shaping strategies, solving problems, embracing change, and achieving results. I have confidence in my ability to overcome obstacles and "get the job done" with limited resources. I enjoy the challenge of organizing and managing multiple tasks.

I would greatly appreciate the opportunity to further discuss how I could contribute to your organization's future challenges as well as its success. I would be glad to meet with you at your convenience. I look forward to hearing from you and may be reached at any of the above telephone numbers or email address.

Thank you for your consideration.

Sincerely,

Gerald L. Crews

Enclosure–Resumé